The Politics of Cyberconflict

T0199664

The Politics of Cyberconflict focuses on the phenomenon of 'cyberconflict' (conflict in computer-mediated environments and the internet) and looks at the way it has impacted on politics, society and culture.

Athina Karatzogianni provides a framework for analysing this new phenomenon, by adopting elements of social movement, conflict and media theory. This new volume theoretically and empirically locates and introduces the key issues confronting global politics today, as a consequence of the impact of new communication technologies.

This new book examines the current technological environment and investigates the topical areas of:

- Power, participation and democracy in internet politics.
- Internet security, information warfare and cyberterrorism.
- Sociopolitical cyberconflicts, such as social movements, dissidents against governments and internet censorship in different countries.
- Ethnoreligious cyberconflicts, such as al-Qaeda and the Israeli–Palestinian conflicts.
- An analysis of the internet's role in the organization of anti-war protests during the 2003 Iraq conflict and its effect on media coverage and the impact of the war on the internet itself.

Students of new media, politics, sociology and conflict studies will find this book to be of great interest.

Athina Karatzogianni's current research focuses on the further implications of information communication technologies for ethnoreligious conflicts, contemporary social movements and the global media.

Routledge research in information technology and society

The Politics of Cyberconflict

Athina Karatzogianni

LONDON AND NEW YORK

First published 2006
by Routledge
2 Park Square, Milton Park, Abingdon, Oxon, OX14 4RN

Simultaneously published in the USA and Canada
by Routledge
270 Madison Ave, New York NY 10016

Routledge is an imprint of the Taylor & Francis Group, an informa business

Transferred to Digital Printing 2008

© 2006 Athina Karatzogianni

Typeset in Times by Wearset Ltd, Boldon, Tyne and Wear

British Library Cataloguing in Publication Data
A catalogue record for this book is available from the British Library

Library of Congress Cataloging in Publication Data
A catalog record for this book has been requested

ISBN10: 0-415-39684-0 (hbk)
ISBN10: 0-415-47980-0 (pbk)
ISBN10: 0-203-96962-6 (ebk)

ISBN13: 978-0-415-39684-4 (hbk)
ISBN13: 978-0-415-47980-6 (pbk)
ISBN13: 978-0-203-96962-5 (ebk)

Contents

Acknowledgements

First and foremost, I am utterly indebted to my family – Magdalene, George and Manos – whose persistence, confidence and character I attempted to breathe into this work.

As this book is based on my doctoral research, I would especially like to thank my supervisor Professor Ian Forbes, for motivating me to pursue my passion for the subject and for showing patience, encouragement and understanding. I am also grateful to the School of Politics and its people at the University of Nottingham, and above all my friends in the A1 research room, for swapping ideas and critiques with me over the past four years. I would also like to thank the *Journal of Politics* and the *Journal of Cultural Technology and Policy* for permission to include early pieces on cyberconflict and the internet and the Iraq war respectively.

Lastly, this work would have been far less enjoyable without a great deal of help and hindrance (!) of virtual, real and imaginary friends, mentors and co-travellers. These include all the authors cited in this book, my editors Harriet Brinton and Heidi Bagtazo and their team at Routledge, as well as Andrew Robinson, Vasilis Lappas, George Michaelides, Christiana Gregoriou, Phoebe Moore, Jonathan Dunne, Petros Ioannidis, Kostas Tzortzis, Matthew Humphrey, Richard Aldrich, Robin Brown, Shanthi Sekaran, Kanellos Terzis and especially Timothy Hawkins. This cannot but include my students and their fresh and intimate way of looking at global politics. This book is dedicated to them.

Introduction

The thrust of this analysis is to disclose the extent to which the internet has affected politics and political conflict in particular (cyberconflict) and provide a theoretical framework for explaining this phenomenon. The aim of the book is to first theoretically and then empirically locate and introduce the new issues confronting global politics today, due to the impact of new communication technologies.

The structure of the book reflects this strategy. The first chapter begins to explore some of the main features of internet politics (social movements, power, participation, democracy, globalization) in the form of a literature review, with the central argument that the political game remains traditional, despite the use of a postmodern medium. Since the literature did not examine political conflicts manifest on the internet, such study was urgently required, in order to come to grips with the consequences of the use of this technology by political actors and groups. This leads to Chapter 2, where, in order to explain the empirical evidence of 'cyberconflict', a proposal was made to integrate the elements of social movement, conflict and media theories into a single analytical framework for cyberconflict. An understanding of cyberconflict requires an examination of the technical environment where this takes place (internet security, information warfare) and its sociopolitical implications. This is achieved in Chapter 3.

Searching for a satisfactory description of empirical cases of cyberconflict led to the use of a classification between two types of cyberconflict: sociopolitical and ethnoreligious. The following three chapters contain examinations of different instances of this classification. Chapter 4 looks at the sociopolitical cyberconflicts: social movements, dissidents against governments and internet censorship in different countries. Chapter 5 examines examples of ethnoreligious cyberconflicts and the possibility of conflict resolution. This is followed by the last empirical Chapter 6, which includes discussion and analysis of the internet's role in the organization of anti-war protests during the Iraq 2003 conflict, its effect on media coverage and the impact of the war on the internet itself. The conclusion of all the empirical chapters singles out which of the variables mentioned in the integrated theoretical framework can be linked to the empirical examples outlined. The Conclusion summarizes the most important arguments and findings this book brings forward.

In terms of methodology, the research is conducted on the following levels:

Textual: Review of internet political literature and main political concepts, social movement, conflict, media theory literatures and information warfare and internet security texts.

Theoretical: The use of the three theories is justified theoretically and the elements considered crucial for cyberconflict analysis are integrated into a single theoretical framework.

Empirical: The study uses more than 400 articles and web sources to locate examples of cyberconflict.

Analytical: Components of the proposed integrated theoretical framework are linked with the empirical evidence while examining these examples, attempting to prove the applicability of it to cyberconflict.

1 How traditional concepts and issues fit into a global postmodern medium

Faces of Israelis and Saudis and Iraqis, detailed charts of military hardware, reports from over decorated generals, from people with strange accents and from points all over the world, a barrage of images and information, packaged in easy-to take portions, punctuated by commercial breaks where powerful fantasies flow by too quickly – the world has truly been faxed, cabled, express delivered.

(Poster 1995: 159)

The field of research undertaken is politics on the internet and, more precisely, how political conflict manifests itself on the internet. In the face of different approaches to internet politics, my aim is to analyse these approaches, indicate what the problems are and outline how my work fits into this background. This chapter addresses the political problem of how new social movements use a postmodern medium like the internet to achieve traditional political goals, such as democracy, power and participation in an era of globalization.

In political terms, the internet is viewed as a vehicle for educating individuals, stimulating citizen participation, measuring public opinion, easing citizen access to government officials, offering a public forum, simplifying voter registration and even facilitating actual voting. It has been termed a powerful technology for grass-roots democracy and one that, by facilitating discussion and collective action by citizens, strengthens democracy. It also has been called potentially the most powerful tool for political organizing in the past 50 years. The underlying purpose of this chapter is to demonstrate that, while the demands of the groups using the internet remain traditional and modern in their characteristics, the medium itself is postmodern, serving modernist ends.

It is important to understand what the central political aims of groups using communication technologies are, why and how they use them and to what effect. With this in mind, this chapter navigates through the literature, building a platform from which to embark on a study of cyberconflict and its meaning for new communication politics.

Political movements and their use of the internet

Social movement theory is particularly relevant to this research, because many of the groups using the internet have social movement characteristics – for example, the Zapatista movement in Mexico. The internet is much cheaper than other broadcasting media, is not sufficiently censored yet to impede its use by diverse groups and has a potential worldwide audience, facilitating a successful new way for social movements to carry their message to a much greater audience. However, the issue is that what these groups struggle for, in the final analysis, are traditional modernist concepts like democracy, participation and, above all, power. These concepts will be examined in that order in the next sections, in order to determine if they remain traditional, when they are linked to a postmodern medium, such as the internet. Characteristically, Daniel Nugent asks of the Zapatista movement: 'How can the EZLN move beyond the politics of modernity when their vocabulary is so patently modernist and their practical organization so emphatically premodern?' (Nugent 1995 quoted in Ronfeldt *et al.* 1998: 113). Taylor and Jordan in their work *Hacktivism* distinguish between two kinds of rights sought after by the Zapatistas:

> The primary demands of the Zapatistas are for health, welfare and citizenship rights … In this struggle, the Internet functions as a medium through which the demands for these rights and the struggles around these rights can be communicated. Information rights appear here as almost a second political order, serving the 'first order' rights to health, welfare and full citizenship.

> (2004: 97)

When basic interaction and networking are primarily conducted via the net, time and space no longer restrain individual engagement. In this respect, the density of a movement's targeted social group, which has been portrayed as the major element in fabricating organized action, has to be redefined. Politics outside governmental boundaries in non-governmental organizations (NGOs) or, more broadly, in social movements will also be restructured by shifts in information commodification. Most commonly discussed in this context is the Zapatista revolt in Mexico, where what initially looked like a guerrilla war quickly became a media war. On the day that the North American Free Trade Agreement (NAFTA), signed by Mexico, Canada and the USA came into effect as a free-trade zone, the Zapatistas occupied an area of the Chiapas region in Mexico. When the national army arrived, they withdrew to the rainforests. They successfully created such a media event that they were able to force the Mexican government into negotiation and avoid a full-scale war. They did this by full use of the new possibilities for information flow, including the internet (Jordan 1999: 166). An exciting quote is the following:

> Meanwhile, the EZLN called on Mexican civil soceiety – not other armed guerillas, but peaceful activists – to join with it in a nationwide struggle for social, economic, and political change, without necessarily taking up arms.

The EZLN also called on international organizations (notably, the Red Cross) and civil actors (notably, human rights groups) to come to Chiapas to monitor the conflict. This was not at all a conventional way to mount the insurrection.

(Ronfeldt *et al.* 1998: 2–3)

Cyberspace offers a medium in which people can interact and coordinate their actions without relying on a face-to-face contact (Tsagarousianou *et al.* 1998: 8). A thriving example of this is the anti-globalization movement whose participants organize heavily through the net. Dissident political groups can now have a voice that is very difficult for governments to silence. These groups have been able to mobilize support, in order to facilitate dialogue between such groups and their governments, such as happened between the Sendero Luminoso (Shining Path) and the Peruvian government (Everard 2000: 158). Everard mentions that the build-up to recent Indonesian elections and the subsequent overthrow of the Suharto regime saw the internet become an active player as Indonesians sought alternative sources of information, while the authorities tightened media controls. The left-wing People's Democratic Party went underground after they were blamed by Indonesian authorities for a riot in Jakarta in July 1996. They continued to respond to accusations on internet discussion lists. Activist groups are increasingly turning to the internet and other electronic media to provide information about their activities and about the activities of their opposition. A similar thing happened in 1997 in Sri Lanka, when the government launched a national website to counter the Tamil Tigers, who have been fighting a long civil war to establish a homeland for Tamils in Sri Lanka. The Tamil Tigers had set up their own website giving their side of the dispute. Sri Lanka officials said their website will 'help in countering anti-Sri Lanka propaganda by enabling Internet surfers worldwide to have access through a single window to authentic news and information on Sri Lanka' (Margolis and Resnick 2000: 20). In the Philippines in the late 1990s, media and political history was made by using SMS (Short Messaging Service) technology or texting in helping bring down former president Estrada.

Nevertheless, such political implications of the internet were not immediately picked up by researchers. Early internet literature focused on issues of hacking, encryption and the use of the internet by extremist groups. It is nonetheless important to mention some of the main concerns expressed, if we are to understand the issues arising from the application of the internet in our everyday lives.

Denning and Baugh commented on the international issue of encryption:

Law enforcement agencies have encountered encrypted email and files in investigations of pedophiles and child pornography, including the FBI's Innocent Images national child pornography investigation. In many cases the subjects were using Pretty Good Privacy to encrypt files and email ... We were told of another case in which a terrorist group that was attacking business and state officials used encryption to conceal their messages.

(2000: 108)

In this context, Barrett makes an important point when he asserts that the internet can support global chains of pyramid letters, anonymous hate mail, offensive graffiti and a range of other 'anti-social' activities:

> Just as fraudulent traders are provided with the potential to operate from 'data havens', so too are political agitators and even terrorists afforded havens from which to criticize government policies – either our own or our allies' – from within the shelter of the UK, protected by anonymous ftp servers, by complicated cross-posting articles, or by the use of dial in access to some other country's machines. In early 1996 for example, plans of the British Army's establishment in Northern Ireland were published on the Internet.
>
> (Barrett 1996: 203)

On the extremists' use of the internet, Rathmell argues that the most remarked upon users of Information and Communication Technologies (ICTs) have been right wing militias in the USA, Islamist opposition movements originating in the Middle East and single-issue pressure groups such as environmental activists or human rights campaigners (Rathmell 2000: 230). However they may differ in aims, membership or ideology, all of these groups have been quick to exploit ICTs for propaganda and psychological operations. Many insurgents, from the provisional IRA through Mexico's Zapatistas to Lebanon's Hizbollah, have incorporated ICTs into their more traditional propaganda and fund-raising activities. They have websites as well as newspapers. Whine writes as follows: 'The German authorities express increasing concern about the Internet. They have noted the growth of far right home pages, from 30 in 1996 to 90 in 1997 and the manner in which football hooligans were mobilized during the 1998 football World Cup series by German Nazis via their websites' (2000: 237).

Also in July 2002 right-wing extremists looking for converts appeared to be trying to subvert the anti-globalization movement, using at least one intentionally confusing website and even showing up at major protests to recruit activists directly. In response to this move the Anti-Defamation League went on the offensive against a site of the 'Anti-Globalism Action Network' (www.g8activist.com), which could easily be accessed by mistake by those intending to visit the G8 Activist Network (Kettman 13 July 2002). In December of the same year, the North Rhine-Westphalia state in Germany required internet providers to block two US-based neo-Nazi websites after a court ruled the measure did not violate the provider's rights. The providers were to appeal to a higher state court and threatened to leave the state for other German states that do not have limiting regulations (Associated Press 20 December 2002). A tough approach on net hate speech has also been taken by the Council of Europe, which has passed a provision which updates the European Convention on Cybercrime, criminalizing internet hate speech, including hyperlinks to pages that contain offensive content. In contrast with the Americans – the council cited a report finding that 2,500 out of 4,000 racist sites were created in the US – many European countries have existing laws outlawing internet racism (Scheeres 9 November 2002). European states have also set up the European Network and Information Security Agency which was set to begin in January 2004 and last until December 2008, with a budget of $28.7 million. The agency would work on problems such as preventing network failures, computer crashes, viruses and unauthorized interception on communications, advising the EU institutions on information security matters (Pruitt 9 October 2003).

Extremist groups use the internet because it is cheap. For the price of a computer and a modem, an extremist can become a player in national and world events. ICTs lower the threshold for participating in illegal acts and, without state or other backing, extremists will look for cost-effective instruments. Furthermore, ICTs act as a force multiplier, enhancing power and enabling extremists to punch above their weight. They can now have a reach and influence that was previously denied to them. Tragically, Helsinki experienced a transfer of activities from the net to real life in late 2002, when a 17-year-old boy in an internet chat room dealing with explosives was held for questioning in a deadly bombing at a suburban shopping mall. The teenager had contacts with a 19-year-old, Peri Gerdt, suspected of making and planting the explosive device that killed him and six others.

To sum up, some political actors are denied access to traditional political means, and ICTs provide new opportunities. The question centres on how these new technological possibilities affect the political situation. With the emergence of this new technology, political actors have unlimited access to easier and cheaper means of political communication. Instead of using traditional means like election campaigns or public relations, the groups that use the internet are able to communicate messages to a wider audience than that reached by more traditional means of political communication. These new technological opportunities affect the political situation in various ways. Political communication becomes more mechanized, it is instant and cheaper and new groups which were previously excluded can take part in a political situation without feeling excluded through the new technology. The groups that use ICTs affect the political situation in that they put forward new rules of the game, the rules of new technology. As a result, traditional political means are less effective and need to adjust to these new technological possibilities. The use of ICTs, and the internet in particular, provide endless opportunities to groups that are otherwise excluded from traditional political communication. This does not mean, however, that new social movements, like anti-globalization, anti-capitalist or anti-war movements, when using the internet to communicate political goals, ask for anything that is not traditionally modern in character, like participation, democracy or power. The nature of the medium does not appear to affect the essentially modernist nature of the game. Lastly, what this first impression points to is that two types of actors use the internet for political purposes: sociopolitical movements and extremist or ethnoreligious groups like the ones mentioned above.

Internet politics: democracy, participation, power

Democracy in internet politics

In order to discuss the connection between democracy and the internet, it would be useful to include a conceptualization of the term 'democracy' before we apply a political theory of democracy to our internet research. The internet brings forth new types of participation in government, thus challenging the traditional type of democracy, where new forms of power configurations can exist between communicating individuals. Arguably, information is the lifeblood of democracy. Keeping this in mind, can we speak of a new politics on the internet or is democracy a traditional term left unaltered by the postmodern realities of the new medium?

In the contemporary world democracy can only be fully sustained by ensuring the accountability of all related and interconnected power systems, from economics to politics. These systems involve agencies and organizations, which form an element of and yet often cut across the territorial boundaries of nation-states. This is how Held views it: 'The possibility of democracy must, accordingly, be linked to an expanding framework of democratic institutions and procedures – to what I have called the cosmopolitan model of democracy' (1995: 267).

Held makes three important points. First, processes of economic, political, legal, military and cultural interconnectedness are changing the nature, scope and capacity of the sovereign state from above, as its regulatory ability is challenged and reduced in some spheres. Second, the way regional and global interconnectedness creates chains of interlocking political decisions and outcomes among states and their citizens, altering the nature and dynamics of national political systems themselves. Third, the way local groups, movements and nationalisms are questioning the nation-state from below as a representative and accountable power system (see also the discussion on social movement theory on pages 53–71 in Chapter 2). Democracy, Held argues, has to come to terms with all three of these developments and their implications for national and international power centres.

In the same context, the hierarchical structure of the state system itself has been disrupted by the emergence of the global economy, the rapid expansion of transnational relations and communication, the economic growth of international organizations and regimes, and the development of transnational movements and actors – all of which challenge its efficacy. With the spread of the internet, there is scope for a newly international localism that is finding expression in 'virtual' communities, with some people going so far as to suggest that a new global cyberstate is forming (Barrett 1996). There are also signs that online communities will offer further dimensions to personal identity within an already complex world. The problem remains that it is extremely doubtful that these changes will ultimately undermine the notion of sovereignty. According to Everard, people still live within a physical location and the idea of cybersovereignty falters, as it fails to think through the place of the body in cyberspace (Everard 2000: 63). However, it can be argued that his view is rather limited, considering more recent challenges facing states' particularly undemocratic ones, which is discussed in the sections 'Chinese dissidents' (pages 128–143) and 'Internet censorship internationally' (pages 143–152) in Chapter 4.

To continue with Held, the cosmopolitan model would seek the creation of an effective transnational legislative and executive, at regional and global levels, bound by and operating within the terms of the basic democratic law (Held 1995: 272). This would involve the creation of regional parliaments (for example, in Latin America and Africa) and the enhancement of the role of such bodies where they already exist (the European Parliament), so that their decisions become recognized in principle as legitimate, independent sources of regional and international regulation. Accordingly, Held writes:

> Those seeking to advance greater equity throughout the world's regions, peaceful dispute settlement and demilitarization, the protection of human rights and fundamental freedom, sustainability across generations, the mutual acknowledgement of cultures, the reciprocal recognition of political and religious identities, and political stability across political institutions are all laying down elements essential to a cosmopolitan democratic community.
>
> (Held 1995: 281)

In conjunction, Held is right to argue that democracy has an appeal as the 'grand' or 'meta-political' narrative in the contemporary world, because it offers a legitimate way of framing and delimiting the competing 'narratives' of the good. This is particularly important, because it holds out the prospect of the constitution of the political good as the democratic good – the pursuit of the 'good life' defined under free and equal conditions of participation.

Commensurate with his argument, Held engages in what he calls 'the democratic thought experiment'. At its core, the democratic thought experiment is concerned to describe the obligations people would accept as necessary for the status to be met as equally free members of their political community. In his own words:

> It is an enquiry which aims to abstract from existing power relations, in order to disclose the fundamental enabling condition for possible political participation and therefore for legitimate rule. It is thus, an analytical mechanism, which helps discriminate among forms of acceptance or compliance to political arrangements and outcomes.
>
> (1995: 161)

But, what is the role of citizens in Held's democracy? Citizens, he says, can enjoy liberty only if the power of the state is circumscribed by law; that is, circumscribed by rules which specify limits on the scope of political activity – limits based on the rights of individuals to develop their own views and tastes to pursue their own ends and to fulfil their own talents and gifts. The stratification of autonomy produced by modern corporate capitalism goes beyond the immediate impact of economic inequalities on the capacities of citizens to participate as equals in their collective associations, for there the very capacity of governments to act in ways that interest groups may legitimately desire is contrained (Held 1995: 246).

According to Held, democratic theory and practice face a major challenge. The business corporation, or multinational bank, exercises a disproportionate 'structural influence' over the polity and, therefore, over the nature of democratic outcomes. Political representatives would find it extremely difficult to carry out the wishes of an electorate committed to reducing the adverse effects on democracy and political equality of corporate capitalism. Democracy is embedded in a socio-economic system that grants a 'privileged position' to certain interests. For these reasons, individuals and interest groups cannot be treated as necessarily equal and the state cannot be regarded as a neutral arbiter among all interests (Held 1995: 247).

Democracy is challenged by powerful sets of economic relations and organizations, which can – by virtue of the bases of their operation – systematically distort democratic processes and outcomes. There is a case that if democracy is to prevail, the key groups and associations of the economy will have to be rearticulated with political institutions, so that they become part of the democratic process – adopting within their very modus operandi a structure of rules, principles and practices compatible with democracy (Held 1995: 251).

Testing Held's hypothesis on this research concern, namely the internet, yields two responses. The first one is that new media increase the scale and speed of information provision and give citizens more control over their information diet. Thereby, they better arm citizens with the information they need in order to participate (Tsagarousianou *et al*. 1998: 6). In this respect, new media help reduce some of the inequalities Held refers to, in the capacities of citizens to participate. As Tsagarousianou mentions, for utopian visionaries, the promise of nearly unlimited information delivered to your monitor in mere seconds is the promise of a better democracy. The internet can help to make all of us more active and more knowledgeable about government, even if 'the state cannot be regarded as a neutral arbiter among all interests' (Held 1995: 247). The second response is the most common one among internet researchers, which states that the internet will not make a great difference, or – as Margolis and Resnick put it – politics will continue as usual because 'in the tradition of American capitalism, the Information Highway is increasingly here to advertise and sell products and services, not to improve the democratic quality of American politics and civic life' (Margolis and Resnick 2000: 73). Margolis and Resnick seem to disagree with potential claims about the power of the internet in relation to democracy. They intimately take the view that

> what has occurred is the normalization of cyberspace. Cyberspace has not become the locus of a new politics that spills out of the computer screen and revitalises citizenship and democracy. If anything ordinary politics and

commercial activity in all their complexity and vitality have invaded and captured cyberspace. Virtual reality has grown to resemble the real world.

(2000: 2)

They back this up by saying that political life on the internet has moved away from cybercommunities, in which civic life centres around free discussion and debate. It has entered an era of organized civil society and structured group pluralism with a relatively passive citizenry. Nevertheless, Margolis and Resnick realize that the web is a wonderful vehicle for citizens who are already interested in politics: 'If they choose, ordinary citizens can acquire a plethora of data and gain access to the public face of the democratic political process. It has functioned well as a tool for citizen awareness campaigns, although so far these have mostly concerned politics that affect the net' (2000: 22).

What they also recognize is that to realize democratic control, it is necessary to restore the balance among branches of government. Computer-mediated communication, they believe, could be the vehicle for this, because, among other functions, it can provide citizens with access to the same documentary information as bureaucrats. Nevertheless, such an account denies the effect of the internet and ICTs on the political process, which is empirically evident in countless ways in more recent communication practices.

Interestingly, the same authors see a way that the internet will help spread democracy; but not because more people will conduct more of their politics online. Rather, information at the centre of new forms of economic growth will facilitate the spread of that wealth:

Areas of the world lacking old-fashioned sources of wealth rooted in natural resources such as vast tracts of arable land or scarce minerals can now compete with nations more abundantly endowed. Because the information spread via the Internet will contribute to general world economic growth and because economic growth facilitates the rise of democratic government, the Internet will contribute to the spread of democracy.

(Margolis and Resnick 2000: 210)

This assertion is dubious for two reasons. First, the authors do not provide evidence for their claim that economic growth facilitates the rise of democratic government, which is problematic, especially in the cases of wealthy Middle Eastern states. Second, there is no guarantee that this new economic wealth will not simply end up in the already rich North.

However, one has to agree with Margolis and Resnick about the difficulty of predicting whether the internet will improve the quality of democracy by creating a more informed citizenry. On the one hand, the internet offers instant and almost cost-free information, which could enable the ordinary citizen to be fully informed about all policy areas and to follow government activity in a far easier and less beraucratic manner. On the other hand, one has to remain sceptical, since the availability of unlimited information might inspire those who are already politically active, but does not ensure that the internet itself will increase the population of the attentive public. Unfortunately, those who are uninterested and remain uninvolved in politics will continue to be so, whether the internet exists or not. Also, as Gerodimos argues, 'the leveling of the playing field can liberate long-oppressed voices, but it can also create an alternative public sphere that suffers from a lack of transparency, accountability and regulation that resembles the deficit of the mass-mediated public sphere' (Gerodimos 2004).

In a discussion of the relationship between the internet and politics, one element of Held's hypothesis appears to be true. This element asserts that democracy is challenged by powerful sets of economic relations and organizations which can – by virtue of the bases of their operation – systematically distort democratic processes and outcomes. The internet does now and will continue to discriminate economically. Internet use requires internet access, which costs money many people cannot pay. As suggested by Richard Davis's evidence regarding internet use in the US, '[p]eople making over $75,000 annually are far more likely to own a computer, have access to the Internet, and use the Internet to visit political sites than those who made under $40,000. In contrast the average income nationally is approximately $32,000' (1999: 182).

Manuel Castells also provides us with evidence to support the notion that the internet is not as democratic as one might think, offering an insight on the social, racial, gender, age and spatial inequality in internet access:

> Worldwide, 30 per cent of Internet users had a university degree, and the proportion increased to 55 per cent in Russia, 67 per cent in Mexico and 90 per cent in China. In Latin America, 90 per cent of Internet users came from upper income groups. In China only 7 per cent of Internet users were women. Age was a major discriminating factor. The average age of Internet users in the US was 36 years, and in the UK and in China was below 30. In Russia, only 15 per cent of Internet users were older than 45. In the US households with income of $75,000 and higher were 20 times more likely to have Internet access than those at the lowest level of income.
>
> (2000: 377)

A more recent EU report titled 'A sustainable eEurope: Can ICT create economic, social and environment value?' (August 2003) found that 88 per cent of all internet users account for only 15 per cent of the world's population in the industrialized world, while countries as digitally advanced as Finland have more internet users than the whole of Latin America. There are also big differences within the developed world. In 2002, some 58 per cent of all Americans had internet access at home, compared with only 38 per cent in Europe. Within the EU the percentage varies between 60 per cent in the Netherlands and 10 per cent in Greece. The report also suggests that not enough research has been done to understand social problems caused by the digital divide (http://europa.eu.int/comm/enterprise/ict/policy/ict-sust/final-report.pdf). Moreover, the future of cyberdemocracy is seen as deferred 'given the heavy handy involvement of untrammeled economic and commercial interests in deciding the fate of technologies putatively valuable from the point of view of extending opportunities for political involvement to a larger segment of US society' (Loader 1997: 113).

In addition, in his work *The Internet and Society*, Slevin anticipated that the increasing influence of big business on the internet brings with it at least two fears for regulators and internet users:

> First, there is concern that if builders of network applications hold a monopoly, they may be in a position to determine license fees for the use of their products. Second, there is the worry that those who build applications and control the interfaces may gain a considerable say over the content of the information communicated and accessed by way of the net.
>
> (2000: 39)

Still, this fear of monopoly and existent communications monopolies, in the fields of computer software or, more traditionally, media (for example Bill Gates or Rupert Murdoch) has not stopped the use of ICTs in remarkable ways. Network technologies are increasingly used in public and political debates and communications, thus promoting dialogue between opposing parties, one of the elements of true democracy. Frissen writes that several 'freenets' have been developed in the Netherlands. They are called 'Digital cities', which use the internet and are organized by government-supported private actors. But the government is also using the internet to organize political debates. In his own words:

> In 1995 we have been running an Internet discussion for the Netherlands' Ministry of Home Affairs regarding a new White Paper on information in public administration. One of the research questions in this project was whether the style and participation in a digital debate differed from ordinary political debates.
>
> (Frissen 1997: 117)

Other crucial elements of a democratic polity are privacy and surveillance, as scrutinized by Raab (1997: 156). The claim that democracy and privacy reinforce each other means that the information-accessibility of democracy is not necessarily achieved at the expense of privacy's information-restriction. The development of accountable democratic institutions and privacy-protecting processes cannot be mutually exclusive occurrences, for each is an important condition of the other. Intruding upon privacy, surveillance may impede democratic liberties, due to its 'chilling' effect on communication or expression. Surveillance thrives in authoritarian regimes that are not exposed to public debate and criticism. ICTs can promote communication, political expression and action amongst citizens who cannot meet in person, provided that they can trust the security of electronic media and their integrity against surveillance. It appears, however, that when participants with unorthodox opinions are being monitored by authorities, then the Information Superhighway's (ISH) potential as a reliable democratic tool is damaged. In this respect, Raab claims that:

> There is much debate, through these channels on the questions of free speech, censorship, pornography, the accessibility of officials and politicians through electronic mail, intellectual property rights, the legal position and liabilities of anonymous remailers, the ability to dispose of secure encryption, and the facilities afforded by secure ISH networks to drug-traffickers and money-launderers.
>
> (1997: 166)

The risk of allowing new technologies like the internet to become rampant features of information exchange and communication produces the fear that push-button democracy will become the norm in the twenty-first century. This may occur less as a formal process of national referenda, and more as a bastardized version of manufactured consent, generated by public relations experts and entrenched political machines (Wilhelm 2000: 139).

To continue, in their work *Cyberpolitics*, Hill and Hughes suggest that for any democracy, information is an essential resource. They point out that one of the major failings of democracy is that so few people take part in the democratic process. In 1996, less than one half of the eligible adult population voted in the US presidential elections, and far fewer are involved in such activities as campaigning, writing letters, or circulating petitions. Hill and Hughes examined both Usenet and chat rooms and came up with various conclusions. One of the important ones is that internet messages about less-democratic nations are far more likely to be anti-government than the messages about more-democratic governments: '[n]ot only is the relationship between country democracy and the probability that a message will be anti-government strong, but it is the only significant, consistent predictor of anti-government statements in the Usenet' (Hill and Hughes 1998: 107).

In the same line of argument, they examined the ideological balance in chat rooms and they found that right-wing threads not only dominate, but that these threads have more chatters and more messages. This might imply that right-wingers are more numerous, more active or both (Hill and Hughes 1998: 129). Or it might reflect the marginality of right-wingers, who feel unrepresented by parties and formal systems. Not surprisingly, they concluded that chat rooms are a difficult format for thoughtful discussion, because of the short line space and the fast pace, which require people to make snap comments, not thoughtful ones. The level of information is low and there is a small amount of issue discussion. Most chat-room conversations focus on the actions of people rather than the government and its role in society. Thus, Hill and Hughes counteract that the Usenet is a better format than chat rooms for political discussions. As they explain:

The Usenet is slow and more thoughtful and so has more potential for deliberation. We saw that Usenet threads tend to be more informative and they are more ideologically balanced. In general, we think electronic political discussion works better in a format as the Usenet, as opposed to a chat room style. It is not perfect, but we think it offers a better hope for deliberative democracy.

(Hill and Hughes 1998: 131)

One gets the same feeling when using similar, more recent facilities like the MSN or ICQ.

In *Cyberdemocracy* Tsagarousianou and others examine several attempts by local authorities to provide access to their citizens electronically. As Tsagarousianou describes, Amsterdam's Digital City, for example, combines a number of different communicative and civic functions, such as deliberation, public information and some degree of support of grass-roots groups. The IperBole project of the commune of Bologna supports citizens' deliberation, public information provision and, to a lesser extent, supports grass-roots groups. The Santa Monica Pen project enables deliberation and public information provision. The Manchester Information City initiative provides economic regeneration-related information dissemination as well as deliberation and civic information provision. In contrast, the official city of Berlin project is geared towards the provision of local authority and local area information, while Network Pericles in Athens has plebiscitary and deliberative aspects, and allows for limited information mainly (Tsagarousianou *et al.* 1998: 169). These efforts by local governments connect very closely to Held's argument that regional and global interconnectedness can alter the nature and dynamics of national political systems, since local initiatives could be seen as a challenge to the national government.

One of the networks, namely, the Santa Monica Public Electronic network (PEN), is one of the longest-running and most innovative experiments in electronic democracy. The PEN system is an exceptionally valuable study, because Santa Monica was one of the first cities to offer its citizens access to an interactive public electronic network. Docter and Dutton have alluded to the problem:

Santa Monica's experience indicates that the Internet and the multimedia revolution will not short-circuit such major dilemmas of democratic participation as public apathy. Most of the public in Santa Monica and other American cities are not very interested in either real or virtual participation. Nevertheless, technological change in the household and government, illustrated by burgeoning interest in the Internet, suggests that electronic participation is likely further to be developed by governments throughout the US.

(Tsagarousianou *et al.* 1998: 145)

Moreover, as Tsagarousianou explains, most, though not all, experiments in electronic democracy share a number of common characteristics. First, the social actors are initiating or participating in them as means of reviving and invigorating democratic politics, which for a variety of reasons is perceived to have lost its appeal and dynamism. Second, they have been local or regional in their character, being related to more or less territorially bound urban and suburban communities. Lastly, they have been based on broadly similar technological infrastructures (Tsagarousianou *et al.* 1998: 168).

Graham Browning, in his work *Electronic Democracy*, is more concerned with the actual effects of the internet on political practice. He sets forth a very good example of the potential political power of the internet. Richard Hartman, a computer software engineer from Spokane, Washington, and his wife Mary helped bring about the surprise defeat of then Speaker of the House Thomas S. Foley through an anti-Foley political action committee they almost entirely organized through the net (Browning 1996: 2). Also, Browning mentions that online activists claim some of the credit for the election in January 1996 of Senator Ron Wyden, D-Oregon, an enthusiastic supporter of free speech online. Wyden won his race by fewer than 18,000 votes, a measurable proportion of them supplied by voters attracted to Wyden's candidacy through his internet appeals, activists believe. Another example cited by Browning occurred in late September and early October 1994. This is how Browning describes it:

> On September 29 of that year the House of Representatives passed a lobbying reform bill that many organizations feared would require them to reveal the names of their members and the amount of fees they paid lobbyists to represent them in Congress. After the vote, the Christian Coalition posted emergency alerts to the Internet and CompuServe, asking members to contact Congress immediately and register their opposition to the bill. In the 24-hour period before the lobbying reform bill reached the Senate on the following Friday, almost 250,000 people responded to those alerts. The bill died on the floor of the Senate.
>
> (1996: 3)

More recently, the phenomenon of the 'blog' has emerged. Short for a weblog, a blog is an online journal that can turn anyone with an internet connection into a mini-media outlet. Blogs are easy to create, cheap to set up and commonplace on the web. They can draw thousands of readers per day and dozens of posted comments, similar to talk radio for the wired. An example of a famous weblog is Salam Pax from Baghdad, mentioned later in the discussion, which influenced the media coverage during the March 2003 Iraq conflict. In relation to democratic politics, there is the example of Oliver Willis, 25, who has a political platform of his own despite the fact that he is not a rich man or a player in democratic politics. He ran a website called oliverwillis.com, where he posted an essay promoting Vermont Governor Howard Dean for democratic nominee, drawing comments from people he had never met. 'Blogs are the harbingers of a new, interactive culture that could potentially change how democracy works, turning voters into active participants rather than passive consumers, limiting the traditional media's role as a gate keeper, and giving the rank-and-file voter unparalleled influence' (Weiss 23 July 2003).

Among the wired, there is a debate over whether blogs are a new form of discourse or simply an endless feedback loop, a self-enclosed circle of political junkies echoing and challenging each other. However, Willis and other bloggers say that their work has a way of spilling into the offline world, as blogs can often focus attention on issues the traditional media ignores. Due to the fact that official websites can be boring, blogs can lend an aura of authenticity to a campaign. During the US presidential primaries of 2003–4, the web was filled with unofficial blogs in support of Howard Dean, John Kerry, Richard Gephardt, Dennis Kucinich and President Bush. There was also a movement to draft retired Army General Wesley Clark for a presidential run, which drew much of its energy from blogs. Nevertheless, the use of blogs is not seen as transforming the way presidential campaigns communicate, according to a report from Johns Hopkins University's CampaignsOnline.org project. The report found that while there has been much media attention concerning campaign blogs, only four of the nine Democratic candidates for president currently utilize blogs as part of their communication strategy (AlwaysOn 18 November 2003).

A stark example of e-democracy and internet politics is MoveOn.org, which played an influential role in the US 2003–4 primaries and in recent years has proven the ability to raise money – $3.2 million in 2000 and $4.1 million in 2002 – with a membership list of 1.4 million activists. The number of volunteers for MoveOn is ten times larger than any existing presidential candidate developed. In the future, a campaign's internet strategy will become central to the campaign:

> As the example of MoveOn is duplicated by other organizations from the national to the local level, the power and the impact of online organizing and campaigning will be greatly magnified. Any politician with sense enough to fill out qualifying papers will understand that the Internet must be central to their campaign and not an afterthought.
> (The Moveon Primary: What it means, politicsonline.com, 26 July 2003)

Dean raised more than $40 million through online donations. In three months, John Kerry brought in almost $50 million from 400,000 online donations. However, both Kerry and President Bush are eager to translate online support into actual votes (Witt 2 July 2004).

To sum up, there are two camps on the question of whether the internet enhances democracy. The optimist view is that the internet increases the scale and speed of information and gives more control over the information diet, so that the promise of unlimited information is the promise of a better democracy; as described here, projects by local authorities to provide access to their citizens electronically can prove to be a tool for a better democracy. Also, examples of awareness campaigns have shown the internet can be a useful way to organize politically. On the other side of the coin, there are the arguments that those who remain uninterested and uninvolved in politics will continue to do so; that the internet discriminates economically with social, gender, race, age and spatial inequality; and that, ultimately, the internet is dominated by economic interests, demonstrating a growing resemblance to the real world. It would be safer to assert that the internet might not yet be ready to take on full responsibility of a democratizing tool; nevertheless, its potential for helping to create a better democracy is undeniable. From the evidence presented so far, it does not seem that democracy is radically influenced by the internet, and it remains a traditional concept unaffected by the postmodern nature of the medium. The internet has provided improved access to the political system for outsiders, and mechanisms for spontaneous expression of public attitudes, presenting a powerful way of harnessing the political moment. Yet, even as they exploit the web, political professionals are somewhat mindful of its anarchic cultural heritage.

Participation in internet politics

Another issue linked to the present research is participation, because the groups that are availed through the internet use it to promote greater participation in their respective political agendas. Nevertheless, participation is not always linked to democracy. Carole Pateman (1970: 2) suggests that the collapse of the Weimar Republic, with its high rates of mass participation, into fascism, and the post-war establishment of totalitarian regimes based on mass participation, albeit participation backed by intimidation and coercion, underlay the tendency for 'participation' to become linked to the concept of totalitarianism rather than that of democracy.

Pateman refers to Dahl's view about the possible dangers inherent in an increase in participation on the part of the ordinary citizen. The lower socio-economic groups are the least politically active and it is also among this group that 'authoritarian' personalities are most frequently found. To the extent that a rise in political activity brought this group into the political arena, the consensus on norms might decline, and hence polyarchy would decline. Therefore, an increase in the existing amount of participation could prove to be dangerous to the stability of the democratic system.

However, this analysis seems to amount to something akin to elitism, and it is quite disturbing to think that lower socio-economic groups should be excluded from the democratic process because they lower the 'quality' of democracy. In relation to the internet, which can potentially involve mass participation in democratic politics, such an argument would be difficult to swallow.

Further, Pateman deems that certain conditions are necessary if the democratic system is to remain stable. The level of participation by the majority should not rise much above the minimum necessary to keep the democratic method (electoral machinery) working; that is, it should remain at about the level that exists at present in the Anglo-American democracies. The fact that non-democratic attitudes are relatively more common among the inactive means that any increase in participation by the apathetic would weaken the consensus on the norms of the democratic method, which is a further necessary condition. In her own words:

> As we have seen, the formulators of the contemporary theory of democracy also regard participation exclusively as a protective device. In their view the 'democratic' nature of the system rests primarily on the form of the national 'institutional arrangements', specifically on the competition of leaders (potential representatives) for votes, so that theorists who hold this view of the role of participation are first and foremost theorists of representative government.
>
> (Pateman 1970: 20)

Again, keeping the level of participation to the minimum necessary to avoid weakening consensus is a technical and cynical view of participatory politics and limits the potential of new ICTs to generate greater participation.

Rousseau, Pateman writes, might be called the theorist par excellence of participation, and an understanding of the nature of the political system that he describes in the *Social Contract* is vital for the theory of participatory democracy. Rousseau's entire political theory hinges on the individual participation of each citizen in political decision-making. And according to his theory, participation is much more than a protective adjunct to a set of institutional arrangements. It also has a psychological effect on its participants, ensuring that there is a continuing interrelationship between the working of institutions and the psychological qualities and attitudes of individuals interacting with them.

The analysis of the operation of Rousseau's participatory system makes two points clear: first, that 'participation' for Rousseau is participation in decision-making; and second, that it is, as in theories of representative government, a way of protecting private interests and ensuring good government. Rousseau's ideal system is designed to develop responsible, individual, social and political action through the effect of the participatory process. Rousseau also suggests that participation has an additional integrative function; it increases the feeling among individual citizens that they belong in their community. More important is the experience of participation in decision-making itself, and the complex totality of results to which it is seen to lead, both for the individual and for the whole political system; this experience attaches the individual to her/his society and is instrumental in developing it into a true community.

Moreover, Pateman believes that the theory of participatory democracy is built round the central assertion that individuals and their institutions cannot be considered in isolation from one another (1970: 42). The existence of representative institutions at the national level is not sufficient for democracy; for maximum participation by all people at that level, socialization, or 'social training' for democracy, must take place in other spheres in order that the necessary individual attitudes and psychological qualities can be developed. The major function of participation in the theory of participatory democracy is therefore an educative one, educative in the very widest sense, including both the

psychological aspect and the gaining of practice in democratic skills and procedures. Pateman states it thus:

> No longer is democratic theory centered on the participation of 'the people',
> on the participation of the ordinary man or the prime virtue of a democratic
> political system seen as the development of politically relevant and neces-
> sary qualities in the ordinary individual; in the contemporary theory of
> democracy it is the participation of the minority elite that is crucial, and the
> non-participation of the apathetic, ordinary man lacking in the feeling of
> political efficacy, that is regarded as the main bulwark against instability.
>
> (Pateman 1970: 104)

The final observation by Pateman to be included is that when the problem of
participation and its role in democratic theory is placed in a wider context than
that provided by the contemporary theory of democracy, and the relevant empirical material is related to the theoretical issues, it becomes clear that neither the
demands for more participation, nor the theory of participatory democracy itself,
are based, as is so frequently claimed, on dangerous illusions or on an outmoded
and unrealistic theoretical foundation (1970: 111). Thus, Pateman suggests that
we can still have a modern, viable theory of democracy, which retains the notion
of participation at its heart.

Having said that, new technology, such as the internet, can be harnessed to
measure citizens' preferences in representative democracies, and may make it
easier for citizens to respond, thus making political participation (access to
information, deliberation, debate and voting) easier, and thereby resolving the
perceived crisis of participation (citizen alienation, abstention and apathy) in
liberal democracies (Tsagarousianou *et al.* 1998: 6). As earlier, Tsagarousianou
notes that, since the early 1990s a number of European local authorities such as
Amsterdam, Bologna, Manchester and civil society actors, rather than central
governments, have been engaged in experiments in electronic democracy, as
have several cities in the UK and the US. The organizers of these initiatives frequently argued that by embracing information and communications technologies
they could resuscitate declining citizen participation and give new vigour to
local politics (Tsagarousianou *et al.* 1998: 168). Margolis and Resnick also
suggest that, for optimists, political participation in cyberspace approximates an
ideal type of communitarian democracy that emphasizes mutuality (2000: 100).
The time needed to become informed about any topic drops substantially when
citizens can employ 'gopher' or 'archie' servers to locate and retrieve desired
information on a vast variety of topics, including matters of public policy that
comprise the formal business of government.

Nevertheless, there is another side to this story. Even if there has been a
tremendous increase in the number of people now on the internet, there is no
indication that their presence will inaugurate a new era in politics. The internet
has become a mass medium, but the numbers tallied by those who estimate the
growth of the internet have not been translated into comparable growth in polit-

ical participation. Providing greater choice and opportunity only solves part of the problem of participation – for example, that related to reducing the cost of involvement – but it does not get to the heart of what motivates citizens to move from the state of disengagement to one of salutary involvement in civic life (Margolis 2000: 207; Wilhelm 2000: 87). Wilhelm writes that, while many communications researchers suggest that anonymity may liberate the individual and equalize participation in a forum where power is asymmetrically distributed, others argue that the individual's isolation coupled with invisible surveillance and hierarchical observation from the outside may lead to the veritable incarceration of the user (2000: 46). Promoting a diversity of voices, while imperative, does not *eo ipso* guarantee deliberation, negotiation and the contestation of viewpoints. Universal access is not sufficient, Wilhelm thinks, for realizing a discursive, democratic polity. Instead, he argues that 'deliberation or critical-rational reflection is a necessary condition of salutary political conversation online, without which digital democracy may follow the lead of "mature" media and fail to meet expectations' (2000: 86).

Another interesting point on the internet's potential for participation is made by Davis (1999). He did a content analysis of three Usenet groups, an analysis of one week's worth of messages between 14 June and 20 June 1997. Davis, similarly to Hill and Hughes (1998), argues that Usenet 'possesses certain disadvantages as a forum for public discussion of political issues. Those include opinion reinforcement, flaming and under-representativeness' (1999: 161). As Davis suggests: 'Usenet political discussion groups tends to favor the loudest and most aggressive individuals. Those who are less aggressive risk rigorous attack and humiliation' (1999: 163). Also, over the past decade there has been consistent support in the literature for the hypothesis that, in the absence of social context cues, the level of uninhibited verbal behaviour in CMC (computer-mediated communication) rises (Collins 1992; Kiesler *et al.* 1984; Smolensky 1990).

Nevertheless, as a two-way mass communication medium that allows users to receive news and information, inasmuch as they participate in information, transmission and public discussion, the internet potentially diffuses power over information dissemination and public debate. The growth of the internet raises the prospect that the traditional groups who dominate American politics will become anachronistic, Davis argues, like organizational dinosaurs in the cyberspace age. In his own words:

> They will be displaced by new organizations or by citizens who no longer need groups at all. They will become mediators in an era of unmediated communication. When citizens will be able to interact with elected representatives without the necessity of intermediary organizations, what then is the role of interest groups?
>
> (1999: 63)

Furthermore, the internet is not a panacea for local apathy or loss of community, nor is it the curse of atomization. Rather, it is a tool that to the extent of its effect is most likely to maintain the status quo. Information at the local level will be more readily available but, again, whether average citizens will be inclined to utilize it is debatable. This is because people will not change in character simply because they have resources at their disposal with which to follow politics closely. Computer-assisted politics will not be radically different from today's politics. The argument here is that the internet is creating social fragmentation and an increase in both intolerance and incivility, as people end up seeing their fellow citizens as stupid or malicious. According to Davis (1999), this is not healthy for democracy, because it encourages people to choose teams, rather than to think issues through.

Nevertheless, the good news is that internet access is not limited to hugely successful commercial gateways. The free-net phenomena, along with the explosion of homespun Bulletin Board (BBS) networks, ensure that access to information need not be 'filtered' by corporate or government interests. Free-nets, such as the Seattle Community Network, are a combination of electronic bulletin board and online services centred around a specific community. They generally operate on a non-profit basis, gathering operating overheads through donations and community support. One item of interest to organizers, both social and otherwise, is that the National Public Telecomputing Network will help any community centre set up a free-net, and will provide assistance in maintaining the computers and databases essential for its use. Lack of technical expertise no longer need be a barrier to modest start-ups such as these. The commonality of interest expressed in this interaction attests to the empowering nature of the medium (March www.interweb.com/nsmnet/docs/march). March quotes Wellman and Gulia who point out that 'the architecture of the Net facilitates weak and strong ties that cut across social milieus, be they interest groups, localities, organizations or nations, so that the cyberlinks between people become social links between groups that otherwise would be socially and physically dispersed' (Wellman and Gulia 1995: 15). However, it is important to note that these virtual connective structures do not necessarily produce collective identity or ideological cleavage among the people involved, and do not always create communities.

Borrowing Fisher and Kling's words, the inhabitants of these structures are a 'transclass grouping of constituencies and cultural identities' (1994: 17) who recognize that 'there is a critical interaction between organizing efforts, national politics, and nationwide social movements' (1994: 223). Building on these ideas, Fisher and Kling state that 'because community problems almost always originate beyond local borders, the ability to effect change depends to a great extent upon building coalitions, alliances, networks, and progressive political parties' (1994: 17).

In synopsis, the evidence of the local authorities' initiative in European cities suggests that the internet has the potential to resolve the perceived crisis of participation in liberal democracies. However, two points should still be noted. First, that the growth of the internet has not translated into comparable growth in political participation; and second, that universal access – if such a thing exists – is not sufficient for realising a discursive democratic polity.

Nevertheless, while this research was taking place, quite a lot seems to have changed. In stark contrast with these arguments and findings stands evidence from the US primary elections, during which a number of voter-action groups sprung up online, showing that the American public could be mobilized to show a strong online presence, although this did not necessarily translate into a 'real' voting presence. A series of democratic voter-action groups, starting with Howard Dean's supporters as well as John Kerry's, hoped to help their candidates win every possible vote to reach the White House. The Dean campaign, with Joe Trippi at the helm, shattered traditional top-down campaign models by allowing its supporters to act on their own. The campaign broke Democratic fund-raising records by collecting $40 million from 280,000 individuals in 2003. Dean also mobilized 163,000 individuals to attend locally organized Meetups and drew millions to the campaign's official and unofficial blogs. Simon Rosenberg, who, as president of the New Democrat Network, qualifies as a Washington insider, says politics have changed for ever:

> 2003 was the first year of the Internet age of political communications, and the end of the broadcast era. From 1960 to 2003, political communications was dominated by the broadcast model where you spoke out and people consumed it as passive consumers. The Internet age is characterized by the medium, which is participatory, it is interactive, it is one-to-one.
>
> (Singel 2 February 2004)

'To me, what this election is about is a test. A test of the questions of whether a spontaneously self-organizing group of activists on the left can outperform the hierarchically organized, centrally directed political organization on the right', said David Lytel of ReDefeatBush.com, a website run by The Committee to Re Deafeat the President. Driving Votes is another online effort that uses the internet to organize Democrats. This group uses its website as an information post where people nationwide can organize voter-registration road trips to various swing states. 'What's been interesting for me' says Lerner, 'is that there are a lot of people out there who want to do something bigger than they have done in the past. What is exciting to me about the Internet is just that it doesn't cost [much] money to make [our] website, and now lots of people are getting motivated, and I feel we will have a voice' (in Terdiman 23 April 2004). Lytel feels that while Republicans have integrated online and offline fundraising and voter-registration efforts, a great divide exists between Democratic postal and electronic camps, with both sides feverishly protecting their turf. Another organizer, Khoe, explains that Run Against Bush has utilized the internet to sign up more than 2,000 members in 40 states. The group gets together people all over the country for literal anti-Bush running events called Jogs against Bush, or Jabs. Members check the website to find their local jabs, come together, and put on Run Against Bush T-shirts. Sites like these can mobilize the party's often lethargic members. 'Our typical member may be somebody who grumbles about the adminstration and about national security and fiscal policy, but they haven't felt comfortable, for whatever reason, doing something about it' said Khoe (in Terdiman 23 April 2004). And the other anti-Bush grass-roots groups are having similar experiences. Democrats were counting on the internet to help them build a base that would push Kerry over the top on election day. 'It appears that this sort of technique seems to be working well with a progressive mind-set or progressive people . . . because of the two way nature of true online grass-roots participation . . . The users are the ones who choose issues . . . It's a very back-and-forth, multiparticipant sort of thing', said ActBlue's Matt DeBergalis (Terdiman 23 April 2004).

A panel in May 2004 at the University of California at Berkeley Graduate School of Journalism concentrated on political activism in online communities. Their assessment was that the web has been particularly effective in engaging politically apathetic people and motivating them to take politics back to the grass-roots level. The Howard Dean campaign successfully attracted millions of grass-roots dollars because it worked counter to traditional campaign methods, said Markos Moulitsas Zuniga, publisher of the Democratic activist blog, Daily Kos (Zetter 5 May 2004). Typically, consultants and political parties try to suppress voter turnout, but now activists are using the web specifically, in the form of activism networks and blogs, to fight back and reengage people, Moulitsas said. When Moulitsas launched his blog Daily Kos (www.dailykos.com) in early 2002, 'it was really difficult to be a progressive liberal. We were essentially nobodies. But people were taking us seriously. They reasoned that if they could build a large community without possessing any credentials, then they could possibly move that community to accomplish change' (Zetter 5 May 2004). Joe Trippi, campaign manager for Howard Dean, contacted Moulitsas and his partner to become consultants. Moulitsas asserted that the campaign grew because people felt they were participating in a cause, rather than following one: 'Blog audiences want to be part of the discussion. They don't want to be told what to do.' Perhaps that is why Moulitas doubts that the internet could get a candidate elected on its own. 'You still need the grass-roots activities. You still need the traditional media' (Zetter 5 May 2004). This is also the opinion of Craig Newmark, of craigslist, who has commented that the challenge for blogs and social-networking sites is to find a way to get the millions of people they attract to go beyond the echo chamber of the internet.

This section has focused on issues of participation and the internet. The following section continues the discussion of what new political movements are looking for when using the internet by looking at the issue of power in internet politics.

Power in internet politics

Power is an important element in internet politics, as in all realms of politics, since internet groups are interested in the accumulation of power. Apart from participation, the point to be highlighted is what they fight for. Particularly, my aim is to focus on the way opposing groups attempt to accumulate enough power to resolve a conflict in a way that suits them, and to promote their own version of the truth. Despite this obvious connection between power and internet politics, the key issues can be complicated.

Steven Lukes, in his work *Power: A Radical View* (1970), identifies three views of power: the one-dimensional, the two-dimensional and the three-dimensional view. First, the one-dimensional view encompasses writers like Dahl, who in his early article *The Concept of Power* describes his intuitive idea of power as follows: A has power over B to the extent that s/he can get B to do something that B would not otherwise do. In short, as Polsby writes, in the pluralist approach an attempt is made to study specific outcomes in order to determine who actually prevails in community decision-making. Thus, the pluralist methodology, Lukes contends, in Merelman's words 'studied actual behaviour, stressed operational definitions and turned up evidence. Most important, it seemed to produce reliable conclusions, which met the canons of science' (Lukes 1970: 13). Thus, Lukes concludes that this first one-dimensional view of power involves a focus on behaviour in the making of decisions, over which there is an observable conflict of 'subjective' interests, observably expressed in policy preferences and revealed by political participation (Lukes 1970: 25). This focus on behaviour and decision-making, on the part of the pluralists, would not be entirely useful when applying a theory of power to internet politics. This is, first, because power can have other, more complicated dimensions, such as ideology, knowledge and social forces; and second, because the internet involves a much more networked form of power relationship, which the pluralist model is insufficient to explain.

The two-dimensional view of power is a critique of the pluralist view. In their critique, Bachrach and Baratz denoted that it is restrictive, and by virtue of that fact gives a misleading and sanguine pluralist picture of American politics (Lukes 1970: 16). Power, they claim, has two faces. Their central point is this: to the extent that a person or group – consciously or unconsciously – creates or reinforces barriers to the public airing of policy conflicts, that person or group has power. A satisfactory analysis of the two-dimensional power involves examining both decision-making and non-decision-making. A decision is a choice among alternative modes of action; a non-decision is a decision that results in suppression of a latent or manifest challenge to the values or interests of the decision-maker. As a result, non-decision-making is a means by which demands for change in the existing allocation of benefits and privileges in the community can be suffocated before they are even voiced. The two-dimensional view of power involves a qualified critique of the behavioural focus of the one-dimensional view. It allows for consideration of the ways in which decision-making could be prevented on potential issues. There is, over these potential issues, an observable conflict of 'subjective' interests, seen to be embodied in express policy preferences and sub-political grievances.

Lukes is criticizing both views:

> The trouble seems to be that both Bachrach and Baratz and the pluralists supposed that because power, as they conceptualise it, only shows up in

cases of actual conflict, it follows that actual conflict is necessary to power. But this is to ignore the crucial point that the most effective and insidious use of power is to prevent such conflict from arising in the first place.

(1970: 23)

So, Lukes offers a third view of power, the three-dimensional view, which involves a thoroughgoing critique of the behavioural focus of the first two views as too individualistic. It also outlines the many ways in which potential issues are kept out of politics, whether through the operation of social forces and institutional practices or through individuals' decisions. Lukes defines the concept of three-dimensional power by saying that A exercises power over B, when A affects B in a manner contrary to B's interests. His suggestion is that the one-dimensional view of power presupposes a liberal conception of interests, the two-dimensional view a reformist conception and the three-dimensional view a radical conception.

Lukes criticizes the one-dimensional view by saying that pluralists, by studying the making of important decisions within the community, were simply taking over and reproducing the bias of the system they were studying. The one-dimensional view of power cannot reveal the less visible ways in which a pluralist system may be biased in favour of certain groups and against others. The two-dimensional view goes some way to revealing this, but it confines itself to studying situations where the mobilization of bias can be attributed to individuals' decisions that have the effect of preventing currently observable grievances, 'covert or overt', from becoming issues within the political process. This would not be an approach to internet conflict, which involves the use of the internet by opposing groups, because the two-dimensional view is restrictive, in the sense that it does not study power as exercised by collectivities. This suggests that Lukes's three-dimensional view of power is a better fit for our research concern, especially since it looks at collectivities:

How is one to identify the process or mechanism of an alleged exercise of power, on the three-dimensional view? ... There are three features distinctive of the three-dimensional view, which pose peculiarly acute problems for the researcher. As I have argued such an exercise may in the first place involve inaction rather than observable action. In the second place it may be unconscious. And in the third place power may be exercised by collectivities, such as groups or institutions.

(1970: 50)

Closely related to Lukes's radical view of power, although much more interestingly elaborated, is Foucault's conception of power, and his concept of power/knowledge. These begin to achieve prominence in his work with the appearance of *Discipline and Punish* (1979) and *The History of Sexuality* (1978). The twin term arises out of Foucault's analyses, as a result of his disciplinary theory, where surveillance enables the structures of domination to operate. To illustrate the notion of disciplinary power and surveillance, Foucault uses Bentham's Panopticon, a circular architectural structure in which cells are arranged around a central viewing tower in such a way as to ensure permanent surveillance, which ensures control of and discipline over the incarcerated bodies. In the two works mentioned above, systems of knowledge are defined as bound up with regimes of power and truth. Systems of power bring forth different types of knowledge. Foucault (1980) thinks that power and knowledge directly imply one another and that there is no power relation without the correlative constitution of a field of knowledge, or knowledge that does not simultaneously presuppose and constitute power relations.

Consequently, recognized forms of knowledge always bring power, and power in turn, justifies the formation of specific kinds of knowledge. The subjects or objects to be known are all integral elements of power/knowledge strategies. Foucault saw these techniques of power undergoing a two-stage development. Initially, they were instituted as means of control or neutralization of dangerous social elements, and evolved into techniques of enhancing the utility and productivity of those subjected to them. The connection, he proposes, between power and knowledge is not just a particular institutional use of knowledge as means to domination. Foucault objects to the very idea of knowledge outside the network of power relations. He proposed these remarks about power and knowledge, first and foremost, to make sense of how the observation and classification of individuals and populations contributed to newly emerging strategies of domination, and how their applications came to constitute knowledge (Foucault 1980). In Poster's understanding of this process, the individual subject is interpellated by the super-panopticon through technologies of power, through the discourse of databases that have very little to do with modern conceptions of rational autonomy. The super-panopticon, as a perfect writing machine, constitutes subjects as decentred from their ideologically determined entity (Poster 1995: 87).

Foucault opposes approaches to knowledge that privilege a 'sovereign subject' anterior to discourse. Discourse becomes meaningful, not because of the individual, but because of the discursive formation, providing subject-positions which the individual can occupy (McNay 1994: 68). In this way, Foucault argues that there cannot exist a subject prior to language. This is an illusion. Consequently, Foucault totally rejects any notion of the subject. His interest lies in analysing the play of discontinuities in the history of discourses. He does not care about the meaning or the truth of a statement but, rather, is concerned with the rules of formation that determine the objects, concepts, operations and options of a particular discourse (Torfing 1999: 90). While Foucault does not recognize the system of dispersion as a tool to distinguish one discursive formation from another, he identifies four sets of rules of formation, such that if one or more is found in the analysis of dispersed statements, then we can identify a particular discursive formation. The rule sets comprise the following: the formation of objects, the formation of enunciative modalities, the formation of strategies and the formation of concepts (McNay 1994: 67). Another principle he brings in is rarefaction, in order to explain why, in a particular era, everything that can be said is never said. This happens, according to Foucault, because of the reproduction of relations of social domination through the control of meaning. Discourses and meaning are the sites of social struggle.

In sum, Foucault's archaeological project is to examine statements by particular societies that make serious claims to truth, and to describe the relation between statements and their historical transformation. One criticism against his archaeological approach is that, while he explains the construction of discrete subject-positions within discourse, he offers no explanation of the social context in which these positions are embedded, and which governs how they are filled, nor does he consider the issues of power that are inevitably vital when considering the social context of discourse (McNay 1994: 84).

In his investigations of madness in the clinical and scientific discourses of *Archaeology of Madness*, Foucault was heavily criticized for treating discourses as autonomous systems of scientific statements. Foucault took his revenge by turning philosophy on its head in his later genealogical accounts of power/knowledge, where discourses are related to non-discursive practices and processes, such as economic and political changes. He now moves away from a mere description of the historical rules of discourse to investigate how social practices shape discourses, and vice versa. His view of power is new. He sees it as not only a negative, but also as a positive force (bio-power), where discipline aims not only to constrain those over whom it is exercised, but also to make use of their capacities (Hindess 1999: 113). In essence, what Foucault did was to articulate a vital aspect of social relations: discourses are products of power. Power relations and scientific discourses mutually constitute one another. However, Foucault does not abandon the archaeological perspective. It becomes an internal element of his genealogical approach with the two models brought together in what he calls problematization. Nevertheless, as Howarth points out:

> We are thus confronted with four main difficulties in Foucault's conception of discourse. These are his failure to formalize satisfactorily his theory of discourse; his inadequate conceptualization of power/resistance; his lack of concrete analyzes of resistances to power; and his inability (or refusal) to examine the 'macro' strategies and outcomes of power/resistance struggles.
>
> (2000: 84)

Like Howarth, Simons comments that Foucault does not attempt to systematically break down the elements of power/knowledge. Rather, his accounts are a delicate entanglement of power and truth: 'Power/Knowledge is a knot that is not meant to be unraveled' (Simons 1995: 27).

But let Foucault explain:

> Nothing is fundamental. That is what is interesting in the analysis of society. That is why nothing irritates me as much as these inquiries – which are by definition metaphysical – on the foundations of power in a society or

the self-institution of a society, etc. These are not fundamental phenomena. There are only reciprocal relations, and the perpetual gaps between intentions in relation to one another.

(as quoted in Moss 1998: 119)

The rise of the social and the atrophy of the political through the internet are making Foucault's writings increasingly relevant to this analysis. Perhaps Foucault's toolkit may be used to confront our electronic world (Boyle 1997). As power has spread and broadly diffused, escaping the confines of states, it is everywhere, very much as Foucault thought. His disciplinary theory and his biopolitics seem to explain the internet as simply another form of social control through the control of information. Elimination of public space, reducing publics to masses of atomized social agents, connects to Foucault's conception of the institution of hierarchical communication in prisons (1979: 238). Moreover, when we accept that power and knowledge are so tightly linked, then the internet could be viewed as a potential source of knowledge and information. It could be viewed as a powerful tool to those who are able to use it or, more importantly, to control it.

The reason I am interested in discourse theory is that in this research project different types of discourses are analysed (ethnoreligious, sociopolitical, media) and discourse analysis is part of the parameters used in my integrated theoretical framework, as shown on pages 86–93 in Chapter 2.

In response to Foucault's writings on discourse, Laclau and Mouffe wrote extensively on discourse theory. In contrast to Foucault, who maintains a distinction between discursive and non-discursive practices, Laclau and Mouffe reject the distinction. They affirm that a) every object is constituted as an object of discourse, insofar as no object is given outside every discursive condition of emergence, and b) that any distinction between what are usually called the linguistic and behavioural aspects of social practice is either an incorrect distinction or ought to find its place as a differentiation within the social production of meaning, which is structured under the form of discursive totalities (Laclau and Mouffe 1985: 107).

Moreover, Laclau and Mouffe introduce four basic categories when they analyse identity: articulation, elements, moments and nodal points. Identity emerges through the articulation or rearticulation of signifying elements, and articulation is 'any practice establishing a relation among elements such that their identity is modified as a result of the articulatory practice' (Howarth 2000: 9). Every discourse is constituted as an attempt to dominate the field of discursivity by expanding signifying chains, which partially fix the meaning of the floating signifier. The privileged discursive points that partially fix meaning within signifying chains are called nodal points or, as in Lacan, *points de capiton*. The nodal point creates and sustains the identity of a certain discourse by constructing a knot of definite meanings. Torfing explains that, according to Žižek, this does not imply that it is simply the 'richest' word, the word in which is condensed all the richness of meaning of the field it 'quilts' (1999: 98). The *point de capiton* is rather the word, which, as a word, on the level of the signifier itself, unifies a given field and constitutes its identity. Discourse is 'the structured totality resulting from this practice'; 'moments are the differential positions' that 'appear articulated within a discourse' when elements are those differences that are 'not discursively articulated' because of the 'floating' character they acquire in periods of social crisis and dislocation (Howarth *et al.* 2000: 7). When Laclau and Mouffe consider the material character of discourse they argue that it cannot be found in the experience of a founding subject and that diverse subject-positions appear dispersed within a discursive formation. 'The practice of articulation, as fixation/dislocation of a system of differences, cannot consist of purely linguistic phenomena; but instead pierce the entire material density of the multifarious institutions, rituals and practices through which a discursive formation is structured' (Laclau and Mouffe 1985: 109).

Also, these two theorists present two useful concepts: equivalence and difference. Equivalence divides social space by condensing meaning around two antagonistic poles; on the contrary, difference weakens a sharp antagonistic polarity, in order to relegate that division to the margins of society. Another significant contribution of Laclau and Mouffe is to the concept of hegemony. After they analyse Gramsci's notion of hegemony, which asserts that political subjects are complex collective wills and that the collective will is a result of the politico-ideological articulation of dispersed and fragmented historical forces, and also other Marxist conceptions, they go on to argue that the two conditions of a hegemonic articulation are the presence of antagonistic forces and the instability of the frontiers that separate them:

> ... in order to speak of hegemony, the articulatory moment is not sufficient. It is also necessary that the articulation should take place through a confrontation with antagonistic articulatory practices – in other words, that the hegemony should emerge in a field criss-crossed by antagonisms and therefore suppose phenomena of equivalence and frontier effects.
>
> (Laclau and Mouffe 1985: 135)

To continue, Mouffe in her own work argues that it is impossible to speak of the social agent as if we are dealing with a unified, homogeneous entity. Rather, the social agent is constituted by an ensemble of 'subject-positions' that can never be totally fixed in a closed system of differences, constructed by a diversity of discourses among which there is no necessary relation but, rather, a constant movement of overdetermination and displacement (Mouffe 2000: 77). She also agrees with Derrida that the constitution of identity is always based on exclusionary practices and the establishment of a violent hierarchy between the resultant two poles. Lastly, in her most recent work, *The Democratic Paradox*, she follows Lacan in depicting discourse as inherently authoritarian, in the sense that only through a master-signifier can a consistent field of meaning emerge. This is how she puts it:

> For Lacan, the status of the master signifier, the signifier of symbolic authority founded only in itself (in its own act of enunciation), is strictly

transcendental: the gesture that 'distorts' a symbolic field, that 'curves' its space by introducing a non-founded violence, is stricto senso correlative of its very establishment. This means that if we were to subtract from a discursive field its distortion, the field could disintegrate, 'de-quilt'.

(Mouffe 2000: 137)

Laclau and Mouffe offer a useful way of thinking about discourse and ideology, especially since it does not demand a strict mode of thinking by way of methodology, so research can be conducted through incorporating the thinking of other writers as well, without falling out of one's own logic. If we accept the proposition that the social agent is constituted by an ensemble of 'subject-positions', it is far easier to understand the subject in such virtual environments, such as those this work addresses.

Following the writers mentioned above, especially in politics and international relations, many writers have used discourse analysis in their work, with interesting results. The reason for this has a lot to do with a point made by Freeden (1996: 113), when he writes that the text as a pattern of words remains an objective constant, whereas all ideologies – because they are constructed from many texts – are in a continuous process of restatement. Bowman, for instance, argues that, particularly in relation to the nation, political identities are discursively constructed. He writes that a wide range of persons and collectivities can identify themselves as constituent parts of the nation without having their readings and their allegiances to it challenged or denied by particular and exclusionary definitions. This unfixity can only be maintained, however, as long as the persistence of the nation is taken for granted; as soon as the nation is discursively posited as endangered, battle lines are drawn and processes of selective exclusion/inclusion are set in play (Laclau 1994: 144). This is particularly relevant to elements in the ethnoreligious cyberconflict component of the proposed theoretical framework.

For example, deploying Laclauian and Žižekian concepts, Renata Salecl uses discourse analysis and a theory of identity politics to account for the way Serbians create exclusionary identities for their enemies. Salecl writes that all images of the enemy are based on specific fantasies:

In Serbian mythology, the Albanians are understood as pure evil, the unimaginable: that which cannot be subjectivised – beings who cannot be made into people because they are so radically Other. The Serbs describe their conflict with the Albanians as a struggle of 'people with no-people' ... The Croats are portrayed as the heirs of Goebbels, that is, as a brutal Ustashi butchers who torment the suffering Serbian nation – a nation whose fate is compared to that of Kurds.

(Salecl in Laclau 1994: 212)

David Campbell, on his part, urges the abandonment of the realist discourse with respect to the Bosnian conflict and consider a range of political options offered by a deconstructive reading of the conflict (Campbell 1998: 34). Vivienne Jabri, writing on discourses on violence, argues that in recognizing the constructive element of language, 'discourse analysis goes some way towards contributing to an understanding of conflict as an exclusionist discourse reifying a singular way of knowing. Discourses which reify ethnonationalist identity assume a uniformity in human experience which denies a pluralism of identities' (1996: 140).

The way Jabri formulates her theory on discourses on violence is very important to the research undertaken here, especially when combined with Campbell's actual analysis of conflicts and Stavrakakis's way of analysing ideologies.

To continue, as Landow makes clear, Michel Foucault conceives of textuality in terms of networks and links. In *Archaeology of Knowledge*, Foucault points out that the frontiers of a book are never clear-cut, because the book 'is caught up in a system of references to other books, other texts, other sentences: it is a node within a network ... a network of references' (Foucault, quoted in Landow 1997: 3–4).

The political significance of CMC lies in its capacity to challenge the existing political hierarchy's monopoly on powerful communications media and perhaps thus revitalize citizen-based democracy. The vision of a citizen-designed, citizen-controlled, worldwide communications network is a version of technological utopianism that could be called the vision of the 'electronic agora' (Margolis and Resnick 2000: 1). Jordan asserts that the politics of cyberspace is strung along the two axes of access to cyberspace and rights within cyberspace (Jordan 1999). Access is a key area of cyberpolitical debate because demographics show that use of cyberspace (up to 1998) was largely confined to a small and privileged section of the offline population. Rights are a key area of cyberpolitics, because the rights of avatars are unclear and subject to revision by offline interests. The politics of online rights has developed chiefly around the areas of censorship, privacy, intellectual property and encryption. To elaborate on the process, cyberspace can alter problems such as the broadening and democratization of decision-making procedures, by removing the constraint of physical presence. Discussions carried on through email, or by asynchronous posting, as used on Usenet, open possible avenues for greater group participation. Cyberspace offers opportunities for breaking down hierarchies within institutions. The global nature of cyberspace is important here, as it only requires one country connected to the net to allow the publication of some information for that information to be let loose in cyberspace. Information restricted in an offline nation-state will then be available in cyberspace, subverting the national boundaries that have helped in the past to control access to information. Jordan puts it this way:

> Cyberspace appears as a place in which individuals can put aside many of the inequalities of offline life, simply because nobody knows if they are 'really' female, old or disabled ... Cyberspace appears to be a place that

undermines the hierarchies of offline life, in which different hierarchies that come to exist depend on the quality of thought and writing – a place where even the destructive behavior that is peculiar to it, flaming, is still only words.

(1999: 87)

Jordan also writes on the role of elites. Particular structures and pressures in cyberspace feed into the creation of technology, according to values and its use or appearance as inert things. These pressures point towards growing control of cyberspace by elites who are defined by their technical expertise, i.e. their ability to alter the 'thingness' of technology that constructs online life. If cyberspace is crucial to areas of offline life, then those that control or manage these areas will understandably seek some reassurance that cyberspace will continue to provide its services. Producers, consumers and others will all, in different ways and through different representatives, endeavour to ensure that the space they depend on is reliable and secured. To do this, governments will legislate cyberspace, corporations will build and rebuild it to their design, politicians will apply it to electioneering and consumers will demand its support.

Another question to be asked is: who uses or possesses the new power offered by the internet? The United States Commerce Department released a report titled 'Falling Through the Net II' (National Telecommuniations and Information Administration 1998) requested by then Vice-President Al Gore. It examined telephone and computer penetration rates to determine who is and who is not connected to the information infrastructure – in other words, who has the power which internet knowledge can provide. The study concluded that the gap between the digital haves and the digital have-nots has been growing. There is a significant gap between key groups. Whites are more than twice as likely to own computers than African-Americans or Latinos. Among the least connected are the rural poor, rural and central city minorities, young households and single-parent female-headed households (Margolis and Resnick 2000: 144).

Manuel Castells, in *The Rise of the Network Society*, stresses that internetrelated firms are, and will increasingly be, in the twenty-first century, at the heart of new information-technology industries. This is how he justifies it:

First, because of their potential dramatic influence on the way business is conducted. An often-cited projection by Forrester Research in 1998 put the extended value of electronic business transactions in 2003 at about $1.3 trillion up from $43 billion in 1998. But, secondly, the Internet industry has also become a major force on its own ground because of its exponential growth in revenue employment, and market capitalization value.

(2000: 149)

As for the impact of the internet industries on the economy as a whole, Castells informs us that in the US, internet-related jobs increased from 1.6 million in the first quarter of 1998 to 2.3 million in the first quarter of 1999. E-commerce represented the fastest growing sector. According to Castells,

> the speed of development of the new industry was without precedent: one-third of 3,400 companies surveyed in 1999 did not exist in 1996. The growth of revenue in Internet industries in 1999 was projected to account for $200 billion – this in contrast to total growth in revenue in the US economy of about $340 billion. By the turn of the century, the Internet economy, and the information technology industries, had become the core of the US economy – not only qualitatively but quantitatively.
>
> (2000: 151)

An example of the potential impact of the internet on the economy is when, in November 1999, the city of Pittsburgh handed down the opportunity of electronic disintermediation by offering $55 billion worth of municipal bonds directly to institutional investors over the internet, thus bypassing Wall Street. As Castells believes, this was the first time municipal bonds were directly sold electronically and the entry of electronic trading into the $13.7 trillion bond market is even likely to affect financial markets (2000: 154). Nevertheless, it could be argued, due to the collapse of e-companies and e-stock that this bubble is bound to burst (again).

To return to political issues, Jordan provides us with a definition of cyberpower (1999: 208). Cyberpower, in his view, is the form of power that structures culture and politics in cyberspace and on the internet. It consists of three inter-related regions: the individual, the social and the imaginary. Cyberpower of the individual comprises avatars, virtual hierarchies and informational space and results in cyberpolitics. Power here appears to be exercised by individuals. Cyberpower of the social is structured by the technopower spiral and the informational space of flows, resulting in the virtual elite. Power, at this point, appears as forms of domination. Cyberpower of the imaginary consists of the utopia and dystopia that make up to the virtual imaginary. Power, in this respect, appears as the constituent of social order. All three regions are needed to map cyberpower in total, and no region is dominant over any other:

> Lyon, on the other hand, argues that the governance of the new social polarization and the new virtual culture is governance founded upon new matrices of power. State power seeps towards individuals – the new global citizens wired up to the Internet who, through the unintended consequences of their actions, are busily forming new patterns of sociality, new virtual communities and thus new bases for power.
>
> (Loader 1997: 44)

In the final analysis, internet politics, like any other politics, are closely related to power, because individuals or groups using the medium are inevitably involved in power relations. The problem is that not all people have the power of knowledge the internet can provide, since the medium as yet is mostly used by white, middle-class males, a trend that is starting to change slowly, as in our non-virtual social sphere. To explain, the internet offers opportunities for breaking down political hierarchies within institutions, while subverting the national boundaries that have helped in the past to control access to information. Social movements, in particular, might still be asking for traditional forms of political power; however, the new power they gain from the internet could be characterized as somewhat non-traditional, in the sense of giving them access to mass audiences they were previously denied.

The globalization–internet connection

Another issue identified in this research is that of globalization. The reason for this is that the internet can be seen as a global phenomenon, very much contemporaneous to globalization, to the extent that, among its other effects, it has a globalizing influence, allowing differing cultures the chance to co-exist. It is necessary, I think, to look briefly at the debate on globalization, in order to place internet politics in their political environment, especially when the internet is seen by many as a tool of this process.

Processes of globalization are challenging the bases of order in profound ways – first, by exacerbating inequalities both with and among states; and second, by eroding the capacity of traditional institutions to manage the new threats. Globalization transforms the processes, the actors and capabilities, and the agenda of world politics, necessitating more effective international institutions of management. In more contemporary terms, institutions need to probe deeply into domestic politics, ensuring compliance with agreements on issues ranging from the environment to trade and arms control. The concept of globalization describes dramatic changes in the transactions and interactions taking place among states, firms and peoples in the world. It describes both an increase in cross-border transactions of goods and services and an increase in the flow of images, ideas, people and behaviour. Woods puts it this way:

> Economistic views treat the process as technologically driven. Yet, globalization has also been driven by deregulation, privatization and political choices made by governments . . . In other words the impact of globalization has been strongly shaped by those with the power to make and enforce the rules of global economy.
>
> (Woods in Held and McGrew 2000: 389)

Globalization is cementing old economic inequalities between 'haves' and 'have-nots' – not just in the sense of having technology or not, but also in the sense of having the capacity to make rules or not. Yet, at the same time, globalization is creating a new set of requirements for regulation and enforcement, which requires the cooperation of the so-called 'have-nots'.

Currently, there is a debate between globalists and sceptics in theoretical terms. Empirical examples of these two camps are found in the discussion of cyberpolitical dimensions in conflicts (Introduction in Held and McGrew 2000). This debate involves fundamental considerations about the nature of world order, both as it is and as it may be. Disagreements can range over at least three separate dimensions: first, the philosophical concern with conceptual and normative tools for analysing world order; second, the empirical-analytical concern with the problems of understanding and explaining world order; and third, the strategic concern, which focuses on an assessment of the feasibility of moving from where we are to where we might like to be (McGrew 2000: 401).

Globalists seek to review the nature and meaning of the modern polity in its global setting. They reject the assumption that one can understand the nature and possibilities of political life by referring primarily to national structures and procedures. The transnational and global scale of contemporary economic and social problems presents, globalists contend, a unique challenge to the modern state. This challenge involves, in the first instance, the recognition of the way globalization generates a serious 'political deficit' – a deficit which encompasses democracy, regulation and justice. Second, re-examining the changing context of the modern state entails recognizing the way globalization stimulates new political energies and forces, which are providing an impetus to the reconfiguration of political power. These include the numerous transnational movements, agencies and NGOs pursuing greater coordination and accountability in regional and global settings. Third, globalists affirm that a shift is and ought to be taking place between political and ethical frameworks based on the national political community and those based on a wider set of considerations.

In stark contrast, sceptics hold that the modern theory of the state presupposes a community, which rightly governs itself. The modern theory of the sovereign democratic state, they contend, upholds the idea of a national community of fate – a community which properly governs itself and determines its own future. For the sceptics, particularly those who subscribe to the communitarian outlook, the values of the community take precedence over all universal requirements.

According to McGrew, globalization presents modern democratic theory with a daunting task: how to reconcile the principle of rule by the people with a world in which power is exercised increasingly on a transnational or even global scale Today, McGrew believes, the fate of democratic communities across the globe is interweaving ever more tightly with patterns of contemporary globalization, with the result that established territorial models of liberal democracy appear increasingly hollow. In his own words:

> A new agenda for democratic theory is called for: one which breaks with conventional accounts of democracy in which the nation-state is conceived as the only proper incubator of democratic political life. Central to this new agenda is a critical enquiry into the necessity, desirability and possibility of 'global democracy' – that is of democracy beyond borders.
>
> (McGrew 2000: 405)

Furthermore, Held argues that contemporary globalization is transforming state power and the nature of political community, but any description of this as a simple loss or diminution of national power distorts what has happened. For although globalization is changing the relationship between states and markets, the change does not occur directly at the expense of states. Held puts it this way:

> States and public authorities initiated many of the fundamental changes – for example, the deregulation of capital in the 1980s and early 1990s. In other spheres of activity as well, states have become central in initiating new kinds of transnational collaboration, from the emergence of different forms of military alliances to the advancement of human rights regimes.
>
> (2000: 421)

Contemporary globalization has contributed to the transformation of the nature and prospects of democratic political community in a number of distinctive ways (Held 2000: 423). First, the locus of effective political power is shared by diverse forces and agencies at national, regional and international levels. Second, the idea of a political community of fate – of a self-determining collectivity – can no longer be meaningfully located within the boundaries of a single nation-state alone, as it could more reasonably be when nation-states were being forged. Third, national sovereignty today, even in regions with intensive overlapping and divided political structures, has not been wholly undermined – far from it. Regarding the internet, it seems that a new form of nation has developed. It is what Barrett calls the 'cybernation' (Barrett 1996). As the internet has been used by more and more people throughout the world, Barrett argues it has carved a unique cultural niche. The initial subculture of internet users evolved a set of acceptable behaviours, a common history and, arguably, a common identity of beliefs: free speech, protection of civil rights, impatience with naive questions from the newly initiated, and so on. So, Barrett suggests that

> a cybernation – a nation whose communication of commonly held beliefs and philosophies is effected by the Internet or similar mechanisms – already exists. This initial subculture has been added to and developed further as ever more people around the world find their way into 'cyberspace', creating an evolving and essentially self-organizing community.
>
> (1996: 15)

Another theme that usually comes up in discussions of globalization is that of world government. The classical argument for world government is that order among states is best established by the same means, whereby it is established among individual men within the state by a supreme authority, which perhaps too freely equates states to individuals. As Bull asserts:

> It is often argued today for example that a world government could best achieve the goal of economic justice for all individual men, or the goal of

sound management of the human environment. The argument against world government has been that, while it may achieve order, it is destructive of liberty or freedom: it infringes the liberties of states and nations and also checks the liberties of individuals who if the world government is tyrannical, cannot seek political asylum under an alternative government.

(Bull in Held and McGrew 2000: 445)

An argument entertained in the area of internet research, in relation to the issue of world government, is that the spread of the internet brings scope for a newly international localism – that is, finding expression in 'virtual' communities, with some people going so far as to suggest that a new global cyberstate is forming. There are also signs that online communities will offer further dimensions to personal identity within an already complex world.

In the course of the ongoing discussion, one of the most prominent new tools of globalization is the internet. The problem connected to the internet in terms of its globalization potential is that of cultural homogenization. Some countries on the receiving end of information are concerned that humanity is becoming homogenized, and that they will lose their cultural identity in the rush to become Anglo-Western. This is one of the main reasons cited by Singapore for enacting its draconian internet censorship legislation in 1996 (Bull in Held and McGrew 2000: 32). Malaysia, too, has noted this aspect of unfettered information flows and has expressed concern that images and lifestyles presented on the internet may offend, for example, Muslim sensibilities. Burma and Cambodia are not noted for their tolerance of opposition political groups. Their governments have both expressed concern at the consequences of open and unfettered access to the internet, as all the while they come under increasing pressure to connect.

Another issue concerning internet and globalization is that the internet, for all its rhetoric of globalization, is conducted mostly in English. Those who want to access the internet must have a good knowledge of English, which presupposes a Western-influenced education. Only elites within developing countries can get that sort of education, which marginalizes others, especially in rural areas. Moreover, few people in the South have access to the internet, and not all speak English, the language of the internet. Moreover, even post-colonially, information about the South is primarily written from information and research produced in the North. As Everard puts it, 'this can lead to erosion of cultural identity, national values and cultural integrity, through what Holderness describes as the homogenization of humanity' (Bull in Held and McGrew 2000: 37). More interestingly, while internet uptake is growing among those outside North America and Western Europe, internet-based communication within and between anti-capitalist movements continues to be dominated by English language users (Wright 2004: 92).

The one area where cyberspace has undoubtedly brought political change is the emergence of a global system that restructures the power of the nation-state. Jordan explains: 'Finances flowing across national borders show little regard for the interests of the nation-state they flow through. Information spreading instantly throughout cyberspace evades controls that are more easily put in place on nationally based, centralized broadcast media' (1999: 162).

An example of this emerged when emails came out of Russia describing the attempted coup against Gorbachev, providing a commentary 'as it happened' that evaded state censorship. Again in Russia, more than a decade later, text messages broadcast on the internet were sent during the notorious Moscow theatre incident, while the audience was held hostage by Chechen autonomist rebels.

Another point that deserves attention is that of cultural imperialism on the net. In 'The Harvard Conference on Internet and Society' several people expressed their views on the subject (Slevin 2000: 208–210). An obvious one came from Anne-Marie Slaughter, a lawyer at Harvard School, who championed the idea that it is not cultural imperialism we should be worried about, but the divide between those who are privileged and online, and the rest of the world. The second came from Izumi Aizu, an intercultural communications specialist in Japan, who maintained that the internet creates a global arena for 'seemingly minor culture . . . that the mass economy cannot pay attention to. In some developing countries people are now jumping into use the net, not only to absorb knowledge and information from the advanced countries, but to share their own with others' (Slevin 2000: 209).

To sum up, the internet is one of the most prominent new tools of a globalized world. Despite the globalizationist rhetoric, there is a divide between those who are privileged and online and the rest of the world, with the appearance of such problems as cultural homogenization and cultural imperialism, since internet traffic at the moment is mostly conducted in English. Moreover, information spreads instantly throughout cyberspace and so evades attempts, even by powerful states, to control it. The fact that the internet is part of the globalization process could be used as a platform to argue that it does go beyond traditional politics of sovereignty and therefore is postmodern in nature.

A postmodern medium?

One reason for dwelling on the question of whether the internet is a postmodern medium is that the answer one gives is likely to determine one's assessment of the implications and effects of the internet on society, politics and culture. Concerning postmodernity and the internet, the central issues identified here are community, identity, discourse and structure. These will be examined in turn, following a rundown of the general approaches and issues on the subject.

The use of ICTs, an intellectual activity, has far-reaching implications for our notions of autonomy, sovereignty and self-determination. The individual in cyberspace is fragmented into databases and networks. As a result of this fragmentation, the individual as a meaningful entity becomes decentred and multiplied. Does this mean that we become postmodern?

The postmodern refers to a condition of disunity and the fragmentation of knowledge. Postmodernism involves an epistemological shift from the perceived wholeness of knowledge to a realization that information is by its very nature fragmented. According to Chapman there are two views taken on the internet. The first is the transmission view, where the internet is viewed as simply a mechanism for the instantaneous distribution of information at the global level, and the second is the view of internet use as ritual, linked to terms such as sharing, participation, association, fellowship and the possession of common faith, rendering the internet more than the transmission of signals and messages (Chapman). Or, as Poster argues, modernity or the mode of production signifies patterned practices that elicit identities as autonomous and rational, whereas postmodernity or the mode of information indicates communication practices which constitute subjects as unstable, multiple and diffuse, where one gets a sense of fragmentation and the decentred self, multiple and conflicting identities and a subverted order. The first view, essentially a modernist perspective, tends to reduce the internet to a hammer: 'In the grand narrative of modernity, the internet is an efficient tool of communication, advancing the goals of its users who are understood as preconstituted instrumental identities' (Poster: www.humanities.uci.edu/mposter/writings/democ.html). In contrast, the second view takes account of the social dynamics of online culture, and cyberspace becomes the realm of pure possibility. Reason (the 'modern' Enlightenment catchword) becomes a space of possibility. Also, postmodernism is a viewpoint that emphasizes the horizontal over the hierarchical, and the internet is a rhizomatic structure:

> Cyberspace is a 'smooth space' whose potential for expansion is infinite. One rides a 'flux' and a 'flow' of information. One wanders. As a holding environment, cyberspace is rhizomatic (Deleuze and Guattari, 1987). It is like a subterranean stem that grows and spreads horizontally. In cyberspace thought is pushed to the level of the formal operational mind where the rules of thought are themselves transcended and where reason, as a consequence, becomes a space of possibility.
>
> (Emery: www.osb.org/aba/aba2000/emery.html)

This shift to a decentralized network of communication makes senders receivers, producers consumers and rulers ruled, upsetting the logic of the first media age. This is why Poster is calling for a poststructuralist analysis of the modes of subject constitution. Such an account would avoid the continued, limiting and exclusive repetition of the logics of modernity, '[f]or the chief characteristics of subjecthood, the resistance of the new media to modernity lie in their complication of subjecthood, their denaturalizing the process of subject formation, their putting into question the interiority of the subject and its coherence' (Poster: www.humanities.uci.edu/mposter/writings/democ.html: 41).

Another characteristic of postmodernism is the simulation of reality. People lose touch with reality and replace it with symbols of reality. Due to the seeming weightlessness of the online environment, the reality of its production is hidden, negating its value and rendering its product an experience (Thiel 2001). What exists in cyberspace, hyper-reality, image saturation and simulacra seem more powerful than the real; and the cyborgian mixing of organic and inorganic, human and machine follows a postmodern logic:

> Just as we represent the hardware of human existence, so we express through these machines a collective consciousness – the software of human existence – a consciousness increasingly fragmented and eclectic – and we call that fundamentally postmodern embodiment, expressed on that representation of the postmodern body – the internet!
>
> (Poster: www.humanities.uci.edu/mposter/writings/democ.html)

Community in cyberspace

Prominent in Jordan's work is the idea that an imagined community exists in cyberspace. Anderson defined a nation as an imagined political community – imagined as both inherently limited and sovereign (Anderson 1991: 6). It is imagined because it is impossible for all members of the community to meet; they must hypothesize their commonality. It is limited because there are always borders and beyond those borders there are other nations. It is sovereign because it creates its own rules within its borders. Finally, it is a community because, regardless of actual inequalities between members of a nation, it is always conceived as a deep 'horizontal comradeship' in which all are equal as members of the nation. Though Anderson uses this definition to explore the nature of the nation state, something cyberspace helps to undermine, a similar community exists in cyberspace. Jordan comments:

> There seems little doubt that alongside the virtual lives that individuals construct and the virtual societies technopower conditions, another layer of cyberpower exists on the fantasies and nightmares that collectively constitute the imagination of cyberspace. Here will be found the common beliefs of individuals who never meet each other that will move them to fight for their cybercommunity, believe in their cybercommunity and even love their cybercommunity.
>
> (1999: 183)

Similarly, Castells asks: 'So, in the end are virtual communities real communities?' (in Jordan 1999: 155). His answer is that virtual communities are communities but not physical ones, and they do not follow the same patterns of communication and interaction as physical communities do. For him, virtual communities are not 'unreal'; they just work in a different plane of reality. They are interpersonal social networks, most of them based on weak ties, highly diversified and specialised, still able to generate reciprocity and support by the dynamics of sustained interaction.

Slevin accepts that technologies like the internet are opening up opportunities for new forms of communication, when he argues that the production and reproduction of social reality is becoming re-embedded in local communal life in ways which were largely unavailable in previous modern settings. He elaborates as follows:

> The possibilities of virtual reality are boosting to the extreme the dynamism of modern everyday life by heightening the process which Giddens describes as tearing 'space away from place by fostering relations between absent others'; 'the severing of time from space' he continues, 'provides a basis for their combination in relation to social activity … This phenomenon serves to open up manifold possibilities of change by breaking free from restraints of local habits and practices'.
>
> (Slevin 2000: 106)

Identity in cyberspace

In order to understand the impact of the internet on the individual, Slevin draws upon the work of Rheingold (1994), who indicates that America has lost a needed sense of social commonality and that, in the face of such loss, 'virtual communities' just happen to fulfil this need. In this fashion, internet users, Slevin argues, tend to appear as what Giddens describes as 'cultural dopes', not as actors who are highly knowledgeable, 'discursively and facitly', about the institutions they produce and reproduce in and through their actions (Slevin 2000: 107). However, in *Between Facts*, Habermas had rejected much of the 'cultural dope' approach to media studies, arguing instead that citizens adopt strategies of interpretation against media messages (Salter 2003: 125).

But what are the direct effects of the internet on individuals? Slevin comments that people who work at home or on the move, with no real place in the organization which they can call their own, may feel alienated. They no longer have a fixed 'place' of work where they can develop shared experiences and a sense of belonging (Slevin 2000: 132).

To illustrate the point that the internet has a negative impact on the individual, Slevin refers to the HomeNet project. The families in the study came from eight diverse neighbourhoods in Pittsburgh, Pennsylvania. According to the study, greater use of the internet was associated with subsequent declines in family communication. Greater use of the internet to communicate was also associated with declines in the size of both the local and the distant social circle. Individuals who made greater use of the internet *al*so reported larger increases in loneliness. Remarkably, the researchers write that greater use of the internet was associated with increased depression and disengagement from real life (Slevin 2000: 167). Nonetheless, I am not sure of the validity of these findings, when brought under more empirical scrutiny.

Moreover, Everard's analysis focuses on identity issues when examining the internet. He cites Turkle, who suggests that the use of multiple identities in cyberspace merely extends the range of selves available, thus making the individual in a sense more complete, and more comfortable to try out a range of points of view. The point is that the modern notion of individuals being unitary is itself an illusion. The self of language and of symbolic order at large is always virtual – a simulation (Slevin 2000: 125). In addition, the internet allows individuals to expand beyond identities based on geographical territories, towards identities based more on the cultural terrain of cyberspace.

On the other hand, the issue of identity is not merely one of philosophical importance, but also one of immense practical importance for the conduct of states – not to mention identifying the lines of flight from the status quo:

> For example, the issue of the 'authentic' author is an aspect not only of intellectual property rights, but also of authentication for business transaction. Am I the purchaser that I am presenting myself as? Are you a genuine business to whom I can submit my credit card number? Will a third party intercept my credit card details? All these are questions of identity.
>
> (Slevin 2000: 159)

Discourse in cyberspace

> At the core of the digital zone, moreover, is rapid access to the global, informational archive that simultaneously delights and overwhelms any user who turns to it only for empirical certainty, historical veracity, and subjective validation. In the zones of the internet and digital installation artwork, the graphic certitude of the factual is entwined in the figural play of the fictional, the privacy of the personal erodes with the interactivity of the social, the patient quietude of reading is interlaced with the jumpy quickness of surfing, and the quasi-religious contemplation of textuality and high art becomes newly energised by the flashy multi-media quacking of art in the electrifying zone of the digital.
>
> (Murray 2004)

In postmodernism we also find a suspicion and an ironic deconstruction of master narratives, and a trust and investment in micropolitics, identity and local political struggles. In information technology, writing/speech distinctions lose some of their meaning. The written word takes on a more immediate nature and begins to function as if it were speech. Writing thus achieves 'transcendence' on the internet, as a third-order simulation of speech. This perspective helps to explain how power relations are constructed through discourse and how ideological work is done. This issue is addressed in Loader's *The Governance of Cyberspace*. In his own words:

> It is precisely these postmodernist little narratives which may be characteristic of discourse in cyberspace. As Poster remarks 'the Internet seems to encourage the proliferation of stories, local narratives, without any totalising gestures and it places senders and addressees in symmetrical relations'.

Moreover, these stories and the performance consolidate the 'social bond' of the Internet community.

(Loader 1997: 8)

Jordan also comments on the effects of cyberpunk culture on cyberspace:

While it is clear that cyberpunk was a movement, its ideas have had a much broader effect than on just science fiction. Two ideas in particular were prefigured in cyberpunk science fiction that have had a lasting effect on cyberspace: the organization of information as virtual spaces and the nature of virtual bodies ... The first is the most significant, because it attempts to directly describe, picture, dissect and understand cyberspace.

(1999: 25)

Cyberspace has been conceptualized as a net, a matrix, a metaverse and universally, as a place constructed out of information. That is why it is not merely a medium easily controlled, but *another* place altogether: 'The performativity of an utterance ... increases proportionally to the amount of information about its referent one has at one's disposal. Thus the growth of power, and its self-legitimization, are now taking the route of data storage and accessibility, and the operativity of information' (Jordan 1999: 37 quotes Lyotard 1984: 47).

Castells brings forward the actual form of language on the internet. He argues:

To some analysts CMC, and particularly email represents the revenge of the written medium, the return to the typographic mind and the recuperation of the constructed, rational discourse. For others, on the contrary, the informality, spontaneity, and anonymity of the medium stimulates what they call a new form of 'orality', expressed by an electronic text.

(Castells, quoted in Jordan 1999: 389)

When electronic communications are a factor in the theorist's understanding of the subject, language is understood as performative and rhetorical, and as an active figuring and positioning of the subject. With the spread of this regime of communications, the subject can only be understood as partially stable, as repeatedly reconfigured at different points of time and space, as non-self-identical and therefore as always partly Other (Poster 1995: 59).

The internet is a loose, decentred network of mainframes and personal computers that sustains global connections among its users. Internet discourse is routinely off-topic, repetitive, inane or obscene. It is not an overstatement to say that an ethic of anarchy, with disparate voices raised in electronic cacophony, often prevails.

Perhaps more so than any other contemporary theorist, Jean Baudrillard provides a provocative direction for 'navigating' this hyperreal terrain. Although he has not addressed worldwide networking and the internet specifically in his writing, his comments on telematics, along with more general critiques of modernity, provide an interesting means for exploring the internet. From a Baudrillardian perspective, this figuration of the internet as a kind of cybernetic terrain works to undermine the symbolic distance between the metaphorical and the real (Nunes 1995: 314). Baudrillard's reading can be used to move internet studies beyond its modern closures. Replacing the one world with multiple possible worlds, the internet ultimately offers both the seductions and subductions of a postmodern world.

The postmodern argument

When Nicholas Negroponte introduces his readers to his digital world, they find that the impact of digital technology is great, and explore what being digital means and how their lives may be enhanced. Negroponte supports the view that even if nobody has a clear idea of who pays what on the internet, it appears to be free to most users. Even if this changes in the future and some rational economic model is laid over the internet, it may cost a penny or two to distribute a million bits to a million people. In Negroponte's view, computing no longer falls within the exclusive realm of military government and big business. It is being channelled directly into the hands of very creative individuals at all levels of society, thereby becoming the means for creative expression in both its use and development. In Negroponte's words:

> The agent of change will be the Internet, both literally and as a model or metaphor. The Internet is interesting not only as a massive and pervasive global network, but also as an example of something that has evolved with no apparent designer in charge, keeping its shape very much like the formation of a flock of ducks. Nobody is the boss, and all the pieces are so far scaling admirably.
>
> (1995: 181)

For Baudrillard, the shift from the real to hyperreal occurs when we move from mere representation to simulation, a movement already existing in our virtual world. According to Baudrillard, the screen represents an example of the 'satellization of the real' by achieving the escape velocity of hyperreality: 'that which was previously mentally projected, which was lived as a metaphor in the terrestrial habitat is from now on projected entirely without metaphor, into the absolute space of simulation' (1988: 16).

Another characteristic of the internet, which gives it a postmodern character, is that it becomes a hyperreal vehicle for travelling across a simulated world. The image of 'cybertravel' offers a metaphorical world beyond a computer screen, a 'globe' of nomads that no longer stands for the world, because it has become the 'world'. The Microsoft question that used to pop on the screen 'Where do you want to go today?' creates the simulation of a form of power and the creation of a virtual world which destroys the conceptual possibility of distance. This is why Baudrilland writes that 'the Telecomputer Man experiences a very special kind of distance which can only be described as unbridgeable by the body ... The screen is merely virtual and hence unbridgeable' (1993: 55).

Moreover, the internet does not simply annihilate distance; it creates its own simulated world in place of the physical world of spatial distances. Once the internet moves closer to total connectivity, this metaphorical 'cyberspace' could become the hyperreal – more real than the real place it once simulated (Nunes 1995: 316). Baudrillard refers to this moment as the 'precession of simulacra', when the globe/model defines the world it once approximated (1983: 2). Carrying this Baudrillardian reading to its further limits, one might conclude that what occurs is 'the end of space through cyberspace, the end of knowledge through information and the end of the imaginary through the hyperreal' (Nunes 1995: 319).

For millions of netters, cyberspace is a real place, more egalitarian than elitist and more decentred than hierarchical. Rheingold asks: '[h]ow are relationships and commitments as we know them even possible in a place where identities are fluid?' (1991: 61) We reduce and encode our identities as words on a screen, decode and unpack the identity of others. The fact that many people believe virtual communities to be real places in which they live real experiences, makes this blurring of the real and the unreal close to Baudrillard's postmodern moment of the hyperreal. The following abstract is telling:

> Cyberspace is not real! Hacking takes place on a screen. Words aren't physical, numbers ... aren't physical ... Computers simulate reality, such as computer games that simulate tank battles or dogfights or spaceships. Simulations are just make-believe, and the stuff in computers is not real. Consider this: If 'hacking' is supposed to be so serious and real-life and dangerous then how come nine-year-old kids have computers and modems? You wouldn't give a nine-year-old his own car, or his own rifle, or his own chainsaw – those things are 'real'.
>
> (Sterling 1992: 84)

Finally, the promise of unlimited information and the threat of this information being controlled leads to the realization that sophistication in technology produces more convincing simulations of information and more convincing strategies of deterrence. With the fascination of unlimited information what we are not realizing is that, despite all this information, we have nothing to learn:

> The addiction we have for the media, the impossibility of doing without them ... is not a result of a desire of culture, communication and information, but of this perversion of truth and falsehood, of this destruction of meaning in the operation of the medium. The desire for a show, the desire for simulation ... is a spontaneous, total resistance to the ultimatum of historical and political reason.
>
> (Poster 1995: 16, quoting Baudrilland 1988: 55)

Nonetheless, it should be noted that Poster does not fail to criticize Baudrillard's work, which remains infused with a sense of the media as unidirectional, and therefore does not 'anticipate the imminent appearance of bidirectional, decentralized media, such as the Internet, with its new opportunities for reconstructing the mechanisms of subject constitution' (1995: 19). However, if we follow Baudrillard all the way, it seems as if the audiences comprise helpless and gullible idiots who are incapable of interpreting or reinterpreting images of violence. Signs are made possible by the new technologies of the media in which signifiers flash past potential consumers. Once signifiers have been separated and abstracted in this way, floating freely in communicative space, so to speak, they can be attached to particular commodities by the arbitrary whims of advertisers. According to Baudrillard, individuals consume meanings rather than products, resulting in symbolic exchange (Poster 1995: 107). As Poster argues:

> Electronically mediated communication opens the prospect of understanding the subject as constituted in historically concrete configurations of discourse and practice ... In turn such a prospect challenges all those discourses and practices that would restrict this process, would fix and stabilise identity, whether these be fascist ones which rely on essentialist theories of race, liberal ones which rely on reason, or socialist ones, which rely on labor.
>
> (1995: 77)

A poststructuralist approach to communication theory analyses the way electronically mediated communication (what Poster calls 'the mode of information') both challenges and reinforces systems of domination that are emerging in a postmodern society and culture: 'the figure of the self, fixed in time and place, capable of exercising cognitive control over surrounding objects may no longer be sustained ... electronic communications systematically remove the fixed points, the grounds, the foundations that were essential to modern theory' (Poster 1995: 60). Continuing from Derrida, Poster writes that both deconstruction and electronic writing understand the volatility of written language, its instability and uncertain authorship. Both see language as affecting a destabilization of the subject, a dispersal of the individual, a fracturing of the illusion of unity and fixity of the self (Poster 1995: 72).

Karim, in 'Diasporas and their communication networks: Exploring the broader context of transnational narrowcasting', addresses the question of the postmodern when he mentions Arjun Appadurai (1996), who sees the global cultural economy as characterized by fundamental disjunctures between what he identifies as five dimensions or 'scapes' of 'global cultural flow': ethnoscapes (people), mediascapes (media content), technoscapes (technology), finanscapes (capital) and ideoscapes (ideologies). The diasporic site becomes the cultural border between the country of origin and the country of residence – Homi Bhabha's 'third space' (Bhabha 1994). Hall views this process as operating 'on the terrain of the global postmodern', which 'is an extremely contradictory space' (Hall 1997). These global networks are allowing for relatively easy connections for members of communities residing on various continents. In opposition to the broadcast model of communication, which apart from offering limited access to minority groups, is linear, hierarchical and capital-intensive, online media allow easier access and are non-linear, largely non-hierarchical and relatively cheap (Karim *et al.* 1998). As governments seek to prevent terrorism by more tightly sealing national borders, transnational movement is becoming problematic for potential emigrants from non-Western states. Additionally, the loyalty of minority ethnic groups living in Western countries is becoming suspect and their transnational connections and relationships are coming under scrutiny. The multiple and hybrid identities of diasporic members are under renewed pressure to conform to the mythical notion of a monolithic populace of the traditional nation-state.

The *homo politicus, le citoyen,* is no longer the dominant actor. Systems are becoming more intelligent and developing their capabilites at a growing pace. As an intellectual technology, ICT has far-reaching implications for our notions of autonomy, sovereignty and self-determination. Reality is the unintended result of decisions which are increasingly taken by machines, and the individual in cyberspace is fragmented in databases and networks. As a result of this fragmentation, the individual as a meaningful entity becomes decentred and multiplied. Thus we become postmodern (Frissen 1997: 125).

In pursuit of similar answers, albeit in a more philosophical mode, Deleuze and Guattari in their work *A Thousand Plateaus* (1987) provide us with a detailed analysis of the subject's multiplicities and dimensions. They formulate the concept of the rhizome, which has the characteristics of connection and heterogeneity: any point of a network of rhizomes can be connected to anything else in the network, and must form such connections. This is very different from the tree or root, which plots a point and fixes an order. The linguistic tree on the Chomskian model still begins at a point S and proceeds by dichotomy. On the contrary, Deleuze and Guattari assert that not every trait in a rhizome is necessarily linked to a linguistic feature; semiotic chains of every nature are connected to very diverse modes of coding (biological, political, economic, etc.) that bring into play not only different regimes of signs but also states of things of differing status. There the subject is no longer a subject, but a rhizome, a Body without Organs. In other words, what these writers argue is that a rhizome ceaselessly establishes connections between semiotic chains and organizations of power (1987: 6–7). Multiplicities cease to have any relation to the One as subject or object, natural or spiritual entity; rather, they are rhizomatic, they are flat, a plane of consistency of multiplicities, defined by the outside:

> by the abstract line, the line of flight or deterritorialization according to which they change in nature and connect with other multiplicities. The line of flight marks: the reality of a finite number of dimensions that the multiplicity effectively fills; the impossibility of a supplementary dimension,

unless the multiplicity is transformed by the line of flight; the possibility and necessity of flattening all of the multiplicities on a single place of consistency or exteriority, regardless of their number of dimensions.

(Deleuze and Guattari 1987: 9)

Deleuze and Guattari use the example of a map to explain the rhizome. What distinguishes the map from the tracing is that it is entirely oriented towards an experimentation in contact with the real. It fosters connections between fields, the removal of blockages on bodies without organs, the maximum opening of bodies without organs onto a plane of consistency. It is itself a part of the rhizome: 'the map is open and connectable in all of its dimensions; it is detachable, reversible, susceptible to constant modification' (1987: 12). This resembles the world of the internet, which features connections between fields, bodies without organs, lines of flight and maps of planes of consistencies.

In contrast, arborescent systems are hierarchical systems with centres of significance and subjectification, controlled by central automata like organized memories. An element only receives information from a higher unit, and only receives a subjective affection along pre-established paths. Deleuze and Guattari point to problems in information and computer science, when these sciences grant all power to a memory or central organ. The writers cite Pierre Rosenstiehl and Jean Petitot: 'accepting the primacy of hierarchical structures amounts to giving arborescent structures privileged status ... In a hierarchical system an individual has only one neighbor, his or her hierarchical superior ... The channels of transmission are preestablished: the arborescent system preexists the individual, who is integrated into it at an alloted place' (Rosenstiehl 1974 quoted in Deleuze and Guattari 1987: 16).

In contrast to these centred systems, the authors set forth acentred systems, finite networks of automata, in which communication runs from any neighbour to any other, channels do not pre-exist, individuals are interchangeable – defined only by their state at a given moment – local operations are coordinated and the final, global result synchronized without a central agency (Deleuze and Guattari 1987: 17). That is also what happens when 'mass' movements or molecular flows are constantly escaping, inventing connections that jump from tree to tree and uproot them: a whole smoothing of space, which in turn reacts back upon striated space (Deleuze and Guattari 1987: 506). Or as Hardt and Negri put it, 'a world that knows no outside. It knows only an inside, a vital and ineluctable participation in the set of social structures, with no possibility of transcending them. This inside is the productive cooperation of mass intellectuality and affective networks, the productivity of postmodern biopolitics' (2000: 413).

This again resembles the way new movements and cultures rely on new communication technologies and specifically the internet:

Little could this philosopher of the rhizome have foreseen the intensification of cross-global identity that has been catalyzed by digital culture. An aspect of Deleuze's notion of collective agency that bears noting, particularly in the context of current international politics, is its global positionality, one balanced always-between South and North, not to mention East and West, a positionality whose decisiveness to gesture no doubt varies depending on one's specific place on the global vector and the digital divide. Indeed the machinery of the digital divide now invades and erodes the beingness of always-between at almost every turn.

(Murray: 2004)

Conclusion

This chapter has discussed how traditional concepts and issues fit into a global postmodern medium. It has viewed early social movements and their use of the internet; addressed the question of whether the internet enhances democracy; analysed how issues of power and participation relate to the internet; examined the globalization–internet connection; and discussed the postmodern nature of the medium.

The section on political movements on the internet provides examples of ethnoreligious and socopolitical cyberconflict, looking at early movements on the internet such as the Zapatistas in Mexico, nationalist opposition groups, right-wing extremist and Islamist opposition movements. The groups that use information communication technologies affect the political situation in that they put forward new rules of the game, the rules of new technology. Nevertheless, this does not mean that new social movements like the anti-globalization, anti-capitalist and anti-war movements, when using the internet to communicate political goals, ask for anything that is not traditionally modern in character, such as power, participation or democracy.

While questioning if the internet enhances democracy, the relationship was examined between David Held's theory of democracy and the research concern of this work: the internet. It was concluded that there are two responses to the question about democracy. The optimist states that the internet, by increasing the scale and speed of information, promises unlimited information, which heralds the promise of a better democracy. Initiatives by local authorities in Europe are said to support this thesis. The pessimistic view is that those uninterested in politics will remain so and that the internet is dominated by economic interests, discriminating on grounds of social, gender, race, age and spatial inequality. Both views have limited applicability. Nevertheless, the internet has provided access to the political system for outsiders. This indicates that it enhances democracy now and can do so in the future.

Two other issues discussed in this chapter are participation and power. As far as participation is concerned, Pateman's theory suggests that we can still have a modern theory of democracy which retains the notion of participation at its heart. The problem of participation, in relation to the internet, emanates from the fact that providing greater choice and opportunity only solves part of the problem of participation, and does not get to the heart of what motivates citizens to move from the state of disengagement to one of salutary involvement in civic life (Margolis 2000: 207; Wilhem 2000: 87). On the other hand, the architecture of the net facilitates weak and strong ties that cut across social milieus, so that the cyberlinks between people become social links between groups that otherwise would be socially and physically dispersed (Wellman and Gulia 1996: 15). An example of participation hitting record numbers has been the Howard Dean web campaign during the US primary elections. In other words, political participation can be enhanced by using the internet.

On the question of power in virtual politics, Lukes's radical view on power and the Foucauldian conception of power/knowledge and disciplinary power have relevance. A discussion of these theories led to a focus on discourse and an attempt to explain how discourse constructs power relations. Internet politics, like any other politics, are embedded in power relations. The internet offers opportunities for breaking down political hierarchies and subverting national boundaries. The new power experienced by social movements, because of the new medium, comes from this access to mass audiences previously denied, either through cost, time, physical or political restraints.

To continue, the consideration of globalization and the internet led to the conclusion that there is a divide between those who are online and the rest of the world, with attendant problems of cultural homogenization and cultural imperialism. Since the internet is part of the globalization process, and patently goes beyond the traditional politics of sovereignty, this finding suggests it could be postmodern in nature.

This last point on the postmodern character of the internet is also taken up when addressing questions of identity and community, discourse and structure in cyberspace. An imagined community can be said to exist in cyberspace with common beliefs of individuals who never meet each other that will move them to fight for, believe in and even love their cybercommunity. An element of the internet, which gives it a postmodern character, is that it becomes a hyperreal vehicle for travelling across a simulated world. Electronic communications systematically remove the fixed points, the grounds and the foundations that were essential to modern theory. Information technology thus has far-reaching implications for our notions of sovereignty, autonomy and self-determination. The individual in cyberspace is fragmented in databases and networks resulting in the 'self' becoming decentred and multiple. Thus, we become postmodern in the conduct of internet politics, while remaining modernist in our ambitions.

Finally, this kind of literature creates various research-related obstacles. In the first place, none of the literature addresses the phenomenon of cyberconflict (political conflict in computer-mediated environments) in a direct way. Instead, it furnishes us with a more general understanding of cyberpower and cyberspace. This work will expand on this background in the existing literature by taking its concerns in a previously unexplored direction. It will not only examine the various groups using the internet in more depth, but also analyse the way in which opposing parties in a conflict use the internet, something missing from this literature. More broadly, the difficulty with this literature is that the internet is a rapidly developing medium. As a result, it can be difficult at times for the literature (and the researcher!) to keep up with it.

There are various examples of this. One is that Negroponte is somewhat outdated when, in 1995, he refers to the internet 'as a formation of a flock of ducks that nobody is the boss' since more recent evidence by Margolis and Resnick suggests there occurs a normalization of cyberspace, which comes to resemble more and more the real world (Margolis and Resnick 2000). Then, in an interview in *Wired* magazine in 2002, Negroponte characterized our digital world as lilies in a pond. A second example is Manuel Castells's use of economic data of the years 1998 and 1999, to state that the internet economy has become the core of the US economy with the Forrester estimation of $1.3 trillion up in 2003 from $43 billion in 1998, an estimation that proved to be wildly optimistic. Keeping in mind that the most serious work on the internet was written in the last ten years and that the medium has dramatically changed during these years, it is natural to expect that some of it might be outdated and that, as the internet grows and evolves, the literature will develop accordingly. For instance, as the literature developed, a shift was evident from the first years (when writers were concerned mainly with explaining more technical and quotidian aspects of the medium) to the past few years, when there was greater emphasis on cultural issues, democracy, political parties and the internet, identity and community, cybercapitalism and social revolution (Mosco and Schiller 2001; Davis *et al.* 2002; Franda 2002; Gibson *et al.* 2003). This trend signifies that, as internet use intensifies, researchers are probing the subject more deeply and widely. Inevitably, therefore, internet politics has also developed in new and unexpected ways. This means that it is not feasible to examine the politics of cyberconflict using just one analytical approach.

The following chapter introduces the three theories deemed relevant when examining the phenomenon of cyberconflict and proposes an integrated theoretical framework. Social movement theory is analysed in order to advance our understanding of sociopolitical cyberconflicts; conflict theory is discussed in relation to ethnoreligious cyberconflicts; and theoretical perspectives on media are included, to shed light on the media-related aspects of these conflicts.

2   The three theories

Social movement theory

This section looks first at the general themes of how the study of social movements evolved, the nature of 'new' social movements, whether they represent a shift in post-industrial society, and the middle-class orientation argument. Then, more analytically, the classical resource mobilization model of mobilizing structures is used, framing processes and the political opportunity structure to analyse how these are affected when new social movements utilize the internet.

The problem with social movement theories, as even a casual reader would immediately notice, is that most analyses make empirical generalizations rather than providing useful analytical concepts (Melucci 1989: 23). The result is that in need of a theory to support their case studies, theorists pick and mix from their predecessors and come up with a theory of their own to explain their empirical findings. This means that an examination of thirty different social movements could easily lead to thirty different explanations of social movements. This is because theorists aim at getting a global explanation, despite the fact that the field of social movements is elusive and there are differences, for instance, between movements (as forms of mass opinion), protest organizations (as forms of social organization) and protest events (as forms of action) (Tarrow 1988: 421–440).

Traditional approaches to collective action usually formulate several fundamental questions: Through which processes do actors construct collective action? How is the unity of the various elements of collective actions produced? Through which processes and relationships do individuals become involved in, or defect from, collective action? (Melucci: 1989: 20) Two forms of explanation recur. As in social theory in general there is the agent/structure divide, so in social movement theory the explanation is usually based either on structural conditions and dysfunctions of the social system, or on the differences in values and the psychological differences between individuals.

This dualistic thinking is highly problematic. As far as collective action is concerned, it has been explained in terms of breakdown/solidarity models as pointed out by Tilly (1978). Breakdown theories view collective action as the result of economic crisis and social disintegration, while solidarity models see collective action as an expression of shared interests within a common structural location. The former reduces collective action to reaction and marginality, the latter fails to explain the transition from social conditions to collective action (Melucci 1989: 21). In a similar vein, the structure/motivation dualism is also problematic, as Webb points out, because within this dualism collective action is either a product of the logic of the system or a result of personal beliefs (1983: 311–331).

The response to this problematic dualism was equally dual, meaning it was different in Europe than it was in the United States, with theorists formulating two different ways to deal with the 1970s social movements. In Europe, for example, Touraine (1985: 749–788) and Habermas (1976) tried to move beyond the structure–agency dichotomy by emphasizing the need for a systemic approach that links new forms of social conflict to post-industrial capitalism, while other theorists emphasized the identity-oriented paradigm. In the United States, theorists focused on resource mobilization (McCarthy and Zald 1973, 1977, 1979; Gamson 1975; Tilly 1978). Their theory analysed how a movement is formed, how it persists through time and how it relates to its environment.

To continue, the Weberian term 'closure' describes the processes by which groups come to be formed through strategies of inclusion and exclusion. The activity of social movements, though they also include processes of closure in this sense, typically represent the opposite side of this process: they are attempts by groups thus excluded to insert themselves into closed groups and into closed processes of negotiation between groups, and by so doing to gain access to new resources and opportunities. New social movements are typically either predominantly movements of the educated middle class, especially the 'new middle class', or of the most educated/privileged section of generally less privileged groups. Scott makes a further point:

> Social movements typically bring about change or attempt to bring it about not by challenging society as a whole, though they may appear to do so, but by opposing specific forms of social closure and exclusion. They do so by thematising issues excluded from normal societal and political decision-making, and by articulating the grievances of groups who are themselves excluded. These two aspects – the exclusion of issues and exclusion of groups – are not separate spheres of social movement activity.
>
> (1990: 150)

In his work *Ideology and the New Social Movements*, Scott considers how social movement networks bind the individual to the movement by creating primary bonds in which interpersonal sanction and commitment can operate. The interpersonal character of these networks minimizes problems of free-riding. More importantly, they create self-identification with the movement. Processes of re-formation of the individual will are simultaneously formations of a collective will. Group identity, not merely individual identity, is formed by the movement at the level of its loose networks. By providing individuals with alternative lifestyles and identities, social movements break down barriers to collective action, challenge 'civil privatism' and substitute values of solidarity for instrumental rationality.

An analysis of social movements, Scott suggests, in terms of social closure and interest intermediation treats the integration of issues and groups into the polity as the criterion of a social movement's success. It thus implies that the continued existence of the movement is not an end in itself, because social movement activity can only be understood in the context of other forms of political expression, both institutionalized and non-institutionalized.

The study of social movements raises various issues. First arises the question: how *new* are the new social movements? 'The question is whether their meaning and place they occupy in the system of social relations can be considered to be the same' (Melucci 1989: 105). Also, Melucci claims that 'those who argue for the newness of social movements have simply mistaken an early phase of movement development for a new historical age of collective action' (quoted in Tarrow 1991: 12–20). The same argument is made by Offe who argues that the values advocated and defended by the NSMs are not new but part and parcel of the dominant modern culture, which makes it difficult to think of movements as flowing either from 'pre-modern' or 'postmodern' subcultures (1985: 848). The interesting question here would be: if the internet is influencing the formation of NSMs, does that qualify them as new types of movement? On the other hand, if NSMs are already established, they are just using the internet as a new way of doing something old. For example, the Independent Media Center (IMC) movement has drawn on a legacy of organizational skills developed by earlier social movements. The IMC was able to surmount some of the barriers of similar earlier efforts by building on the experience of earlier networks, inviting many of the activists from the independent video, community radio and open source movements to participate very early on in the planning, fund-raising and gathering of production equipment. 'The four-hundred-strong crew also used all the old and new media from pens to laptops, and from inexpensive audio tape and cam-recorders to the latest in digital recording technologies' (Kidd in McCaughey and Ayers 2003: 61). The IMC is now a collective of over 150 independent media outlets around the world, with hundreds of volunteer journalists offering grass-roots non-corporate coverage. Their success is not just due to new communication technologies, but also stems from the fact that they carefully built a relationship with social movement activists rather than distancing themselves from political organizing. A similar argument is made by Tormey who writes that the internet did not create anti-capitalist activism; rather, anti-capitalist activities used the internet as a means of connecting to others (2004: 159).

The question also emerges as to whether NSMs are a product of a shift to a *post-industrial* economy? This is debatable, as 'cultural revolutionary activities of (NSMs) can be articulated with very different politico-ideological formations, social groups and classes' (Olofsson 1988: 15–34). The mere presence of NSMs in non-Western nations provides evidence against both hypotheses of state intrusion (a change to a post-industrial economy) as well as the hypotheses of value changes discussed below since the new 'post-materialist' values are theorized as a product of the economic and physical security of a country's population which also is not characteristic of Latin American nations, for example (Richardo 1997 411–430). What seems unique according to Richardo is their ideological (identity) characteristic that seems to break from the past (1997: 425). Nevertheless, new social and political movements use the internet for protests against the post-industrial economy. The anti-World Bank protests used the internet mainly for internal and external communication, education and mobilization, sharing activist resources and discussing logistical matters such as transportation and provisions at protest sites (Vegh 2003: 85). The anti-globalization movement has made its voice heard at major demonstrations, for instance in Geneva (WTO, 5/98), Birmingham (G8, 5/98), Cologne (G8, 6/99), Seattle (WTO, 11/99), Davos (WEF, 9/00), Washington (WB/IMF, 4/00), Melbourne (WEF, 9/00), Prague (WB/IMF, 9/00), Quebec (FTAA, 4/00) and Genoa (G7, 7/01) (Vegh 2003: 88).

To continue, there is the new *middle-class* orientation of NSMs argument, which maintains that new social movements are an outgrowth of this new social class (Klandermans 1994). Eder writes that the struggle to overcome a fear of the non-realization of universal moral concepts such as justice, peace or the good life is the reason for collective protest where 'the petit bourgeoisie fills a role which it has rehearsed throughout history: it plays the role of the guardian of the moral virtues of modernity, a role which it has learned how to play since its birth' (Eder 1985: 889). In addition, an explanation based on value-shift is given by Jenkins, who argues that the middle-class 'participation revolution' was rooted in the shift towards post-materialist values, emphasizing self-fulfilment, supporting demands for direct participation and moral concern for the plight of others. 'When elites challenged these values by manipulative acts and outright rejection, the middle class rallied around the movements' (Jenkins 1983: 535).

Such theories involve an exaggeration of the levels of dissent found in the new middle class (Brint 1984: 30–71). Kriesi, looking at new social movements in the Netherlands, found that the thesis of middle-class support for NSMs can be too narrow, because he found the working class to have unexpectedly strong support for these movements; and too broad, because only a part of the new middle class seems to support them (1989: 1111). The part of this new middle class that seems more likely to participate is that which consists of social and cultural specialists. New middle-class radicals choose the welfare and creative professions, because these professions provide a kind of sanctuary, where they are able to escape direct implication in capitalist economic relations (Kriesi 1989: 1084, referring to Parkin 1968). There is also a conflict inside this new middle class as explained by Kriesi in discussions of his findings in the Netherlands; there is 'an opposition of interests between on the one hand, the technocrats in private enterprises and public bureaucracies who try to manage their organizations, and on the other hand the specialists who try to defend their own and their client's relative autonomy against the intervention of the technostructure' (1989: 1078).

However, an emphasis on the new middle class is not entirely inappropriate. When looking specifically at who uses the internet, statistics point to the middle classes of industrialized nations. According to the United Nations, industrialized nations account for 15 per cent of the world's 6 billion people, 88 per cent of whom are internet users, while 80 per cent of the world's population has yet to make a phone call. More people use the internet in London than in all of Africa and there are more users in South Africa than in all African countries combined (Lebert 2003: 224).

Consequently, for the purposes of this work and because it would be difficult to adopt a single position on the issue, arguments will be drawn from writers belonging to different schools. It goes without saying that this approach will not claim to have found a solution to the problem of theorizing social movements or aim at a global explanation.

Scott contends that an adequate theory of social movements would have to recognize the problematic and effortful nature of mobilization and the consequent organizational constraints. This task has been tackled by theories of 'resource mobilization'. At the same time, resource mobilization theory (RMT) has generated a sophisticated analysis of the mechanics of collective action, of the barriers to it, and of the conditions under which it can operate. Nevertheless, while RMT does build an understanding of the organizational dilemma facing social movements, it is limited by its continued adherence to economic models of human agency, and says little about the content and context of social movements.

RMT, like other forms of political realism, sketches both the limits of political action and the largely instrumental and self-interested nature of that action. Resource mobilization theorists have identified the following types of imperative placed upon organizations and movements by the limitations of collective action: (i) the necessity of providing divisible private benefits, as well as indivisible collective ones, places high organizational costs upon collective bodies; (ii) the search for resources such as external funding therefore becomes a major organizational preoccupation; (iii) at the same time, the organization is restricted in the demands and sacrifices it can realistically expect of participants in its costly and high-risk activities. Scott makes an additional point:

> In stressing the continuity, rather than discontinuity, between social movements, parties and institutionalised forms of political action such as pressure groups and parties, RMT has developed a more plausible account of social movement development. That is to say, it is one which corresponds more closely to the typical development of social movements rather than counterfactually trying to explain the 'failure' of movements to retain their movement character.
>
> (1990: 115)

RMT has had a huge impact on social movement theories. It demonstrated that in order to orchestrate collective action, sophisticated organizational forms and modes of communication are needed. As summarized by Cohen, RMT theorists share the following assumptions: (1) social movements must be understood in terms of a conflict model of collective action; (2) there is no fundamental difference between institutional and non-institutional action; (3) both forms of action entail conflicts of interest built into institutionalized power relations; (4) collective action involves the rational pursuit of interests of groups; (5) goals and grievances are permanent products of power relations and cannot account for the formation of movements; (6) this depends instead on changes in resources, organizations and opportunities for collective action; (7) success is evidenced by the recognition of the group as a political actor or by increased material benefits; and (8) mobilization involves large-scale, special-purpose, bureaucratic, formal organizations (1985: 675).

Critically assessing RMT, one could argue that it sets out a narrow and critically impoverished interpretation of human motivation, which reduces it to instrumental rationality. There are two more aspects of RMT which make it vulnerable to general sociological criticism: first, its decontextualized understanding of preferences, choices and actions; and second, the rigidity of the means/ends distinction it employs.

As Melucci and others argue, RMT-based theories do have problems:

> Structural theories, based on system analysis, explain why but not how a movement is established and survives; they hypothesize potential conflict without accounting for concrete collective action. By contrast, resource mobilization models regard such action as mere data and fail to examine its meaning and orientation. In this instance the how but not the why of collective action is emphasized.
>
> (Melucci 1989: 21)

> In other words it looks as if resource mobilization could be defined independently from the nature of the goals and the social relations of the actor, as if all actors are finally led by a logic of economic rationality.
>
> (Touraine 1985: 769)

For the purpose of this analysis, because social groups use the internet as a resource, we will follow RMT in referring to three broad sets of factors in analysing the emergence and developments of social movements: (1) the structure of political opportunities and constraints confronting the movement; (2) the forms of organization (informal as well as formal) available to activists; and (3) the collective processes of interpretation, attribution and social construction that mediate between opportunity and action (McAdam 1996: 2). The focus will be on the mobilizing structures (the network-style structure of movements using the internet, participation, recruitment, tactics, goals), framing processes (issues, strategy, identity, the effect of the internet on these processes) and the media (and the internet particularly) as a component of the political opportunity structure.

Mobilizing structures

The classical RMT model is used to understand, as McCarthy puts it:

> how mobilising structural forms emerge and evolve; how they are chosen, combined, and adapted by social movement activists; and how they differently affect particular movements as well as movement cycle trajectories. The concepts of political opportunity and strategic framing are, I believe, particularly useful in illuminating these processes.
>
> (1996: 141)

Furthermore, according to Tarrow, people engage in contentious politics:

> when patterns of political opportunities and constraints change and then, by strategically employing a repertoire of collective action, create new opportunities, which are used by others in widening cycles of contention. When their struggles revolve around broad cleavages in society, when they bring people together around inherited cultural symbols, and when they can build on or construct dense social networks and connective structures, then these episodes of contention result in sustained interactions with opponents – specifically, in social movements.
>
> (1998: 19)

In a very restrictive sense, a social movement consists of two kinds of components: (1) networks of groups and organizations prepared to mobilize for protest actions to promote (or resist) social change (which is the ultimate goal of social movements); and (2) individuals who attend protest activities or contribute resources without necessarily being attached to movement groups or organizations.

One of the most important components of resource mobilization theory is mobilization. Mobilization is the process of creating movement structures and preparing and carrying out protest actions, which are visible movement 'products' addressed to actors and publics outside the movement (Rucht in McAdam 1996: 186). It implies a process by which an actor augments its resources through gaining the support of other actors or a process by which those that have not taken an active part in politics are drawn into it (Brown 2000: 2). As Brown explains:

> In evaluating the impact of mobilizations it is necessary to distinguish three levels of outcome. Firstly, the success or failure in gaining the support of potential supporters, secondly, success or failure in modifying the position of those that the mobilization is targeted against and thirdly, the impact on the prospects for future mobilizations.
>
> (2000: 5)

In terms of the mobilizational structure, NSMs are open, decentralized, non-hierarchical and ideal for internet communication. Melucci characterizes the new social movements as

> segmented, polycephalic structures. The movement is composed by diverse, autonomous units that expend an important part of their resources on internal solidarity. A network of communication and exchange keeps the cells in contact with each other. Information, persons and models of behaviour circulate in the network, moving from one unit to another and thus promoting a certain homogeneity of the whole structure. Leadership is not concentrated but diffuse . . .
>
> (Melucci 1989: 14)

We assume that the internet is used particularly by movements with two kinds of structure: (a) informal networks with a large geographical reach, and (b) big, powerful and more centralized social movement organizations. Moreover, the internet appears to play an especially crucial role in issue-focused, transnational campaigns (Van de Donk 2004: 18). Using again the Independent Media Center example, the IMC network is based on a non-hierarchical structure that relies on highly complex processes of networked consensus:

> International meetings are held online. There are a wide array of listserv discussion groups that range from general discussions to finances to translation and technical issues. Meetings are conducted through highly complex processes of decision-making, using a consensus model drawn from the direct action wing of the anti-globalization movement.
>
> (Vegh in McCaughey 2003)

According to Tormey, the internet represents an activism based on networks of self-avowed minorities, rather than on classic models of political organization. Those who wish to 'speak' can do so unmediated by the needs and interests of a perhaps distant leadership intent on sending the 'right' signals to the electorate, powerful states or global institutions. As Tormey puts it:

> Networks can be *extended indefinitely* and in more than one or two dimensions. There is no 'membership' as such, just engagement. There is no brake, organizational, fiscal, or ideological on joining in – merely access to the network. Networks facilitate *temporary alliances, coalitions, agreements, events, interactions*. A network consists of chains of allegiance and intersection or what are sometimes called 'nodal points'; where there is convergence for the purpose of acting in support of some group or cause.
>
> (2004: 159)

But is it unthinkable that political parties use ICTs to redesign themselves into a more 'social movement' type of organization, or that social movements are using them to compete with political parties? ICTs might not profoundly change the very 'logic' of collective action, but they seem to change, in any case, the structure of political communication and mobilization (Van de Donk *et al.* 2004: 5). Political organizations that are older, larger, resource-rich and strategically linked to party and government politics may rely on internet-based communications mostly to amplify and reduce the costs of pre-existing communication routines. On the other hand, newer, resource-poor organizations that tend to reject conventional politics may be defined in important ways by their internet presence (Bennett 2004: 125). ICTs can be effectively used to build and maintain powerful and centralized organizations, but empirical evidence also suggests that ICTs can be effective tools to establish and run decentralized networks that allow those who are technically linked to air their views, and, if needed, to mobilize a virtual or physical community of activists (Van de Donk *et al.* 2004: 9).

A striking example of ICTs facilitating and enabling communities is the open source (http://www.opensource.org) or free software movement (http://www.fsf.org) (depending on how you ideologically approach egoless/gift programming). These communities are building on each other's code, software and applications with remarkable results that can be used freely and improved upon by anyone (see, for instance, Raymond 2001). The networked environment, through which these communities operate, enables the development of technology that competes with multinational corporations like Microsoft threatening their monopoly in the industry. Recently, Michaelides (2006) showed that open-source communities are highly adaptive complex systems. Through the evolving principle of self-organization that local interactions give rise to global phenomena, it was shown that these communities separate into core and peripheral developers. In turn, the emerging two-tier organizational structure enables communities to compensate for the coordination challenges that manifest in networked environments. It would be interesting to see in the future whether a similar model applies to other social movements. It could be that a similar process leads to the emergence of leadership and the manifestation of power dynamics in other networked or even traditional social movements.

Some websites are also products of spectacular alliances between NGOs, culture-jammers, small groups of activists, opinion leaders, or just ordinary citizens with the skills and credibility to succeed in the attention game on the world wide web. 'This could be called network politics – a process in which people, organizations, and groups are included not because of formal status, but because they have specific resources needed in the process' (Rosenkrands 2004: 75). In addition, conventional print media produced by the transnational movements (TMs), for example newsletters, newspapers, and magazines, appear to be of less importance than for the NSMs. Instead, the use of electronic communication, in particular the internet, plays a crucial role for TMs and, to a growing extent, contemporary NSMs as well (Rucht 2004: 50).

Furthermore, the cycle of protests argument states that NSMs are simply recent manifestations of a cyclical pattern of social movements (Tarrow 1983).

Some link the cycles to anti-modern or romantic-ideological reactions to the contradictory and alienating effects of modern societies, others to recurring waves of cultural criticism linked to changes in the cultural climate or to political and social events. Movements change in response to shifts in local or national political opportunities, available resources, the actions of counter-movements, changes in the strategy deployed by states and public opinion (Tarrow 1994; Kriesi 1995; Tilly 1995). Minkoff argues that 'trajectories of protest cycles are jointly determined by increases in the rates of protest and increases in the density of social movement organization' (1997: 780). It may be argued that the internet accelerates protest circles and generally makes it far easier than it used to be to organize protests.

Moreover, key issues to focus on when analysing mobilization structures in NSMs are ideology, goals, tactics, participants, recruitment, entry, movement phase, influence, self-label, distinguishing characteristics, key issues and key organizations:

> New social movements call into question the structures of representative democracies that limit citizen input and participation in governance, instead advocating direct democracy, self-help groups, and cooperative styles of social organization. Taken together the values of NSMs center on autonomy and identity.
>
> (Offe 1985: 817–868)

Issues of participation and recruitment are very important and three points should be taken into consideration as outlined by Snow, Zurcher and Olson: the fewer and weaker the social ties to alternative networks, the greater the structural availability for movement participation; the greater the availability for participation, the greater the probability of accepting recruitment invitation; and, movements which are linked to other groups expand at a more rapid rate than more isolated and closed movements (Snow *et al.* 1980: 790–797). This has been clearly observed with the anti-globalization movement and the peace movement. The NSMs have taken to making good use of the greater participation offered on the internet, as well as the numerous links, cyberlinks or otherwise, translated into social links and ties. The NSMs use the internet to support external activity, they may work within the internet to create a foundation for their activities and they have attempted to influence policy affecting the internet (Salter 2003: 129).

There is also a distinction between defensive and offensive *types* of movements. Contemporary movements combine features of both. They are defensive in that they 'defend spaces for the creations of new identities and solidarities' and offensive in that 'they involve conflict between social adversaries over the control of a social field' (Cohen 1985: 689). Here it could be argued that groups in ethnoreligious cyberconflicts represent a more offensive movement type, in contrast to the more mixed (defensive and offensive) character of groups in sociopolitical cyberconflicts, as will become evident in subsequent chapters.

In a very interesting theoretical synthesis, Touraine moves on to two analytical levels, looking at the structural and cultural dimensions of contemporary society and the conflictual processes of identity-formation of collective actors. He identifies these key elements: pursuit of collective interests, reconstruction of a social, cultural and political identity, changing the rules of the game, defence of a status or privileges, social control of the main cultural patterns, creation of a new order (revolution), national conflicts and neo-communitarianism (Touraine 1981: 751). Following Touraine, it could be argued here that groups in sociopolitical cyberconflicts pursue collective interests which relate to the reconstruction of a social, cultural and political identity. Evidence of this is found in the use of the internet by groups in the anti-globalization and anti-war movements. By contrast, in ethnoreligious cyberconflicts, the element of national conflict is far more relevant.

As far as tactics (or modes of action-internal/external) are concerned, these include mobilizing supporters, neutralizing supporters and/or transforming mass and elite publics into sympathizers. Briefly, tactics are influenced internally by organizational competition and cooperation and externally by public opinion and the state. Since societies provide the infrastructure for movement industries, the development of tactics depends on the affluence, degree of access to institutional centers, pre-existing networks and occupational growth of particular movements (McCarthy *et al.* 1977: 1217). Also important are movement participants' professional or communicative skills, which enable them to participate in the process of identity-building. Evidently, information transmitted by successful protest mobilizations becomes the key indicator of political opportunity for emerging movements and drives the development of broad-based protest cycles (Tarrow 1991: 59). On the internet, social movement groups are able to communicate, to generate information and to distribute this information cheaply and effectively, allowing response and feedback:

> This is in large part because of [the internet's] structure as a decentred, textual communications system, the content of which has traditionally been provided by users. Again, such characteristics accord with the requisite features of NSMs: nonhierarchical, open protocols; open communication; and self-generating information and identities.
>
> (Salter 2003: 129)

It is equally important to note here, as Bennett does, that the effects on social movements are due not so much to the internet as to the network structures established through it: 'uses of the internet may have important effects on organizational structures, both inside member organizations and in terms of overall network stability and capacity' (2004: 136). Studies have highlighted two functions of the net: first, it helps communication in terms of information dissemination, formal networking, and action coordination; second, it helps in building a collective identity among participants and potential participants of the movement (Nip 2004: 233). This raises the broader question: have the new forms of communication changed the 'logic of collective action' or just the speed of protest diffusion? (Van Aelst 2004: 121)

The internet delivers significant services to movements: information dissemination and information retrieval, recruitment, mobilization, soliciting opinions, opinion polling, discussion, facilitating contacts between the organization's members, service, networking, communication and coordination with other organizations (Le Grignou and Patou 2004: 187). As Le Grignou and Patou argue, the internet is an important tool in terms of the diffusion of protest and the consistency of protest, in order to achieve a 'consensual mobilization' (2004: 171). Thus, the internet contributes to three different elements that establish movement formation: a shared definition of the problem as a basis for collective identity, actual mobilization of participants and the construction of a network of different organizations (Van Aelst 2004: 99). For example, during the 1999 anti-WTO campaign, while groups with local ties concentrated on mobilization and direct action, more transnationally based groups provided information and frames to feed the action (Van Aelst 2004: 101).

Mobilization is arguably one of the crucial elements in the movements' organization, incorporating the impact of technology on these movements. Brown emphasizes its importance as such: '[i]ndeed it may be more profitable to analyze the impact of information and communication technologies (ICTs) in terms of mobilization rather than in the more sweeping terms offered by ideas such as cyberpolitik or netwar' (2000: 2). Two key arguments support his conviction. These include the argument that mobilization cuts across the distinction between material and ideational factors, and the argument that mobilization keeps the technological issues firmly embedded in a social and political context (Brown 2002: 15).

Not surprisingly, for instance, websites are action mobilizers. According to Edwards' research on the Dutch women's movement online, the emphasis is first on external information provision. The 'first generation' websites contain the basic (static) information about the organization. Then, organizations expand their websites so as to include more information: background information on the problem area that they address, as well as dynamic information about their activities. Subsequently, organizations develop their ambitions further, and become more focused, using network technology for internal communication purposes. Next, organizations start to develop more advanced interactive functions of the internet in their communication with the environment (Edwards 2004: 194). However, as van Aelst and Walgrave argue, most sites offer the 'basics', such as feedback possibility or a newsletter, mostly via email. More sophisticated ways of interaction and debate such as forums or chat-rooms are limited (Van Aelst 2004: 113). An illuminating aspect of their argument is that the concept of mobilization should be extended 'from (former) "unconventional" street actions such as demonstrations and sit-ins to new virtual actions varying from an online petition to pinning down the enemy's server' (Van Aelst 2004: 114). Costanza-Chock, a community arts activist, has a similar opinion: 'you have all these people collectively mobilizing, engaging in action together, telling their friends, discussing what's happening, taking heart that they're not alone in what they feel is a struggle against injustice. So you have the movement-building elements' (Taylor 2004: 153).

The conclusion is that the internet is reshaping the organizational infrastructure of the movements that use it, in at least three ways: to mobilize resources, to maintain relations with the environment and to manage frames. However, as Edwards argues, in relation to the Dutch women's movement, 'the impact is most visible in the movement's increased capacity for mobilizing resources. To a lesser extent, there are also effects in the management of frames … However, the interactive functions on the websites of organizations in the physical domain are still in their infancy' (Edwards 2004: 200). Next, the effect of the internet on the framing process will be considered.

The framing process: identity, issue, strategy

Frames are the specific metaphors, symbolic representations and cognitive cues used to render or cast behaviour and events in an evaluative mode and to suggest alternative modes of action. Symbols, frames and ideologies are created and changed in the process of contestation (Zald 1996: 262).

The management of frames is a crucial element of collective action and mobilization: 'Building a movement around strong ties of collective identity, whether inherited or constructed, does much of the work that would normally fall to organization; but it cannot do the work of mobilization, which depends on framing identities so that they will lead to action, alliances, interaction' (Tarrow 1998: 119).

Similarly, Doug McAdam argues that the concept of framing is an important and a necessary corrective to broader structural theories, which often depict social movements as the inevitable byproducts of expanding political opportunities (political process), emerging system-level contradictions or dislocations (some version of new social movement theory) or newly available resources (resource mobilization) (1996: 339).

If we consider the concept of 'repertoires of collective action' introduced by Tarrow, action is shaped by and coordinated through the development of those models or scripts shared within a particular society at a particular historical juncture. As a resource, the presence of shared goals or models facilitates mobilization (Clemens 1996: 211). This leads us to what the framing process involves: (1) the cultural tool kits available to would-be insurgents; (2) the strategic framing efforts of movement groups; (3) the frame contests between movement and other collective actors – principally the state and countermovement groups; (4) the structure and role of the media in mediating such contests; and (5) the cultural impact of the movement in modifying the available toolkit (McAdam *et al.* 1996: 19).

More specifically, through 'frame bridging', 'frame amplification' and 'frame extension', movements link existing cultural frames to a particular issue or problem, clarify and invigorate a frame that bears on a particular issue or problem, and expand the boundaries of a movement's primary framework to encompass broader interests or points of view. The most ambitious strategy is the fourth, 'frame transformation', a framing device for movements that seek substantial social change (Tarrow 1998: 110, cites Snow and Benford 1992: 467–476).

According to Zald, social movements draw on the cultural stock of how to protest and how to organize. Templates of organization include skills and technology of communication (e.g. writing newsletters, running meetings), of fund raising, of running an office, of recruiting members, and so on. Repertoires of contention include building barricades, organizing marches, non-violent disruption, and the like. Templates of organization may be drawn from the whole society, while repertoires of contention are available from the whole social movements sector (Zald 1996: 267).

Here, a dilemma emerging is the strategy or identity one as identified by Cohen. In 'Strategy or Identity?' he cites Touraine's argument that the exclusive orientations to identity and to strategy are opposite sides of the same coin, in that they both look at social conflicts in terms of the response to long-term changes (modernization) rather than in relation to the social structure. From such a standpoint, 'society' is stratified in terms of the actor's ability (power and privilege) to adapt to change successfully (elites), her or his success in securing protection from change (operatives) or her or his victimization by change (marginalized masses) (Cohen 1985: 675).

In terms of issues, NSMs engage with such issues as quality of life, redistribution, opposition to the present forms of social life and issues that challenge modern state domination. The value-shift hypothesis states that NSMs stress issues of identity, participation and quality of life (referred to as 'post-materialist' concerns) rather than economic matters. This, however, may not necessarily be true, as the anti-globalization and anti-capitalist movement have shown. The internet connection here is that the rise of technology could possibly have effects which alter the agendas adopted by social movements. One example is that maybe there is a link of the globality of computer networks with the globality of protest, resulting in a fusion of disparate issues. As Kahn and Kellner argue:

> Thus, while emergent mobile technology provides yet another impetus towards experimental identity construction and politics, such networking also links diverse communities such as labor, feminist, ecological, peace, and various anti-capitalist groups, providing the basis for a new politics for alliance and solidarity to overcome the limitations of postmodern identity politics.
>
> (1996: 6)

According to Whittier and Taylor, collective identity consists of three related processes: delineation of group boundaries, construction of an oppositional consciousness or interpretative frameworks for understanding the world in a political light and politicization of everyday life (1992: 104–129). Another aspect of the process is formulating cognitive frameworks concerning the goals, means and environment of action, activating relationships among the actors and making emotional investments which enable individuals to recognize themselves in each other (Waterman 2001: 35). Also, the category of collective interest requires prior analysis of 'what counts as collective advantage and how collective interests are recognized, interpreted, and able to command loyalty and commitment' as Cohen (1985: 685) argues and that it is incumbent on the theorist to look into the processes by which collective actors create the identities and solidarities they defend (1985: 690). This means that the logic of collective interaction entails something other than strategic or instrumental rationality. For instance, Pizzorno, who uses a pure identity model, argues that cost–benefit calculations cannot explain the collective action of new groups seeking identity, autonomy and recognition. An extended version of his claim would be to say that 'collective actors strive to create a group identity within a general social identity whose interpretation they contest' (Cohen 1985: 694). New social movements use sharp antinomies such as yes/no, them/us, victory and defeat, now or never, to build up their identities. This hardly allows for political exchange or gradualist tactics. Even more striking is that they do not rely for their self-identification on either the established political codes (left/right, liberal/conservative) or on socio-economic codes (working class/middle class, poor/wealthy, rural/urban, etc.) (Offe 1985: 830).

An interesting study on collective identity by McKenna and Bargh discusses the idea that Usenet groups provide a place for marginalized persons to communicate with others, thus increasing each person's self-esteem (McKenna and Bargh 1998: 681–694). Their theory hypothesizes that group membership is incorporated into the self, so that the individual will feel himself to be a member of the group. As Myers (2001) explains, their conclusion is that virtual group identities are just as important to the self as face-to-face group participation, and that respondents felt for and identified with the people within the Usenet group, 'to identify common themes, opportunities and potential drawbacks in the integration of ICTs into the communication repertoire of social movements' (Wright 2004).

Moreover, Touraine argues, NSMs have an increasingly temporary and symbolic function, fighting for symbolic and cultural stakes and for a different meaning and orientation of social action (Touraine 1985: 798–800). This is especially true in collective actions taken in cyberspace where symbolic change is the key function. This is why the concept of movement itself becomes increasingly inadequate and one has to give attention to Touraine's preference to speak of movement networks or movement areas, as the network of groups and individuals sharing a conflictual culture and a collective identity. These networks allow multiple memberships, and personal involvement and effective solidarity are conditions for participation. This, however, is 'not a temporary phenomenon, but a morphological shift in the structure of collective action' (Touraine 1985: 800, citing Gerlach and Hine's work *People, Power and Change* 1970). Furthermore, the form of the movement itself is a message, a symbolic challenge to the dominant patterns. The structure of NSMs outlined previously by Melucci and Touraine is the basis for internal collective identity, but also for a symbolic confrontation with the system. 'It makes apparatuses to produce justifications, it pushes them to reveal their logic and the weakness of their reasons. It makes power visible' (Touraine 1985: 813).

Due to the fact that frames are transmitted and reframed by the mass media, and consequently the internet, it has been argued that 'the internet can function as a new medium to expose frames and problem definitions and as a space to create shared meaning and identities among the membership and the constituency' (Edwards 2004: 189). Pini, Brown and Previte, for instance, note how new configurations in computer-mediated communication lead to new patterns and possibilities and foster new coalitions of ideas/identities/frames that challenge existing ones in the 'real' world (Van de Donk *et al.* 2004: 24). This introduces the next discussion, the triangle between framing, political opportunity structure, the media and particularly the internet.

The political opportunity structure: the media – the internet

Any attempt at examining this relationship should include a discussion of the issues of media sensitivity and event density (Snyder and Kelly 1977: 105–123). The press is more likely to report protest events that are more violent, involve more people and persist longer. The key event characteristics they identify are size, violence and duration. These factors are collectively designated as event density. Moreover, researchers using the media as a main source of information have to rely largely on practitioners from another profession, such as reporters, editors and publishers, who make the decisions on which version of experience will ultimately be available for research purposes (Mueller 1997: 821). Normally the media provide information on the actors who disagree, but much less information on what they disagree about (Klandermans and Goslinga 1996: 336).

Also, it is widely known that movements have to walk the fine line between extreme forms of action, which alienate third parties but secure coverage, and conventionality, which (even if potentially persuasive) is ignored by the media. In effect, radical reform groups must master the art of simultaneously playing to a variety of publics, threatening opponents, and pressuring the state, all the while appearing non-threatening and sympathetic to the media and other publics (McAdam 1996: 344). This has been a major concern for groups in sociopolitical cyberconflicts, where symbolic hacking, while drawing media attention, nevertheless deprives the other side of means of expression and invites counter-response. Social movements and mass media have several features in common: they are engaged in a struggle for attention; they want to maximize their outreach; they are confronted, though to different degrees, with competitors. Nevertheless, they not only follow a different functional logic, but also have a strikingly asymmetrical relationship when dealing with each other. This becomes clear when we consider the structural positions of the movements offering conflict, spectacle, surprise and threat, on the one hand, and the media (potentially) granting coverage, importance and sympathy, on the other hand. In a nutshell, this asymmetry stems from the fact that most movements need the media, but the media seldom need the movements (Rucht 2004).

Kielbowicz and Scherer indicate that the media are instrumental for social movements in at least three different ways: (1) media are important means of reaching the general public, to acquire approval and to mobilize potential participants; (2) media can link movements with other political and social actors; and (3) media can provide psychological support for social movements (quoted in Klandermans and Goslinga 1996: 319).

The media spotlight validates the movement as an important player. This suggests that the opening and closing of media access and attention is a crucial element in defining political opportunity for movements (Gamson and Meyer 1996: 285). The media form a component of the political opportunity structure with both structural and dynamic elements. The media system's openness to social movements is itself an important element of political opportunity. On the one hand, the media play an important role in the construction of meaning and the reproduction of culture. On the other hand, the media are also a site or arena in which symbolic contests are carried out among competing sponsors of meaning, including movements (Gamson and Meyer 1996: 287). The role of the media apparatus as a validator for the larger society, whose views need to be taken seriously, makes it a crucial target for a movement's efforts to open political space. The media system operates to favour extra-institutional actors in some ways and institutional actors in others (McAdam *et al.* 1996: 289).

In addition, Waterman uses the term communication internationalism to characterize the terrain in which contemporary movements operate (2001: 215). He outlines four characteristics of communication internationalism: (i) it operates in the field of ideas, information and images, revealing that which is globally concerned and new meanings for it; (ii) it is particularly effective on the terrain of communication, media and culture; (iii) the basic relational principle is that of the network rather than the organization; (iv) the movement needs to be primarily understood in communicational/cultural rather than political/organizational terms.

Interestingly, contemporary movements combine forms of action that impact upon different levels of the social system, pursue diverse goals and belong to different phases of development of a system, or to different historical systems (Waterman 2001: 43). An example of this communication internationalism is what Howard Rheingold describes in his recent work *Smart Mobs: The Next Social Revolution* when he writes about 'smart mobs' (defined as 'people who are able to act in concert even if they don't know each other') reshaping the way societies organize and interact. His thesis is that the mobile internet provides far more than a version of the wired workstation. It is, he asserts, creating a quiet revolution. As Rheingold puts it, '[m]obile communications and pervasive computing technologies, together with social contracts that were never possible before, are already beginning to change the way people meet, mate, work, war, buy, sell, govern and create' (quoted in Glasner 2002). Van de Donk *et al.* are also supporting this view. They claim that 'the internet is not used as a mere supplement to traditional media, it also offers new, innovative opportunities for mobilizing and organizing individuals. The new technologies, however, do not determine these innovations. The internet provokes innovation, but this innovation has to be organized and disseminated' (2004: 6).

The most interesting question is whether traditional media have been sidelined by the alternative media of the internet, a topic to be explored further in the section 'The internet's effect on media coverage' (pages 178–187) in Chapter 6. For example, ICTs could improve a movement's capacity to act in a coordinated and coherent way, to react more quickly to an external challenge, and to become less dependent on the established mass media in conveying their messages to a broader audience (Van de Donk 2004: 11). However, Rucht argues that the net has relativized, but has not replaced, the traditional means of both internal and external communication of the movements (2004: 53).

The effect of the internet (as part of the political opportunity structure) on the framing process is, in fact, a striking one. In terms of framing, the groups that use the internet have been innovative in issues both of identity and strategy, and also in framing the issues themselves. As le Grignou and Patou argue, 'the internet makes visible the fragmented plurality of its action by listing together subjects and causes . . . [I]t simultaneously makes homogenous and coherent a set of analyses, activities and movements which otherwise be scattered' (2004: 172). The emergence of global justice movements that closely link a number of issues such as human rights, social rights, poverty and environmental issues has been greatly facilitated by the use of ICTs.

In terms of identity, Nip found that the participants on the Queer Sisters bulletin board developed a sense of solidarity with the Queer Sisters and shared a culture of opposition to the dominant order, but they fell short of harbouring a collective consciousness (2004: 255). This could perhaps be explained by the difficulty in building collective trust in cyberspace. As Wright argues, 'ICTs have sometimes played a dramatic role in communicating rich, multiple impressions of particular events as they unfold, but they have been used less successfully in promoting a coherent, collective assessment of what these events mean within the overall process of social change' (2004: 90).

At the final analysis, the internet could improve a movement's capacity to act in a coordinated and coherent way, to react more quickly to an external challenge and to become less dependent on the established mass media in conveying its messages to a broader audience. The most extreme version of this argument is Diebert's, which stresses that the role of the internet went beyond facilitating activism already in place; rather it helped create 'a new formation on the world political landscape' (Nip 2004: 233).

Another significant factor in social movement theory is system uncertainty. Those who govern have not only to deal with institutional systems of representation, but also with new forms of action such as those used by social movements. The resulting conflicts cannot easily be adapted to the existing channels of participation and representation (e.g. transformed into political parties). In addition, social movements can test the limits of a system; they can 'violate the boundaries of a system, thereby pushing the system beyond the range of variations that it can tolerate without altering its structure' (Waterman 2001: 29). This system uncertainty is increased by the new tactics possible through the internet, which either directly use computers (such as hacking) or help social movements to organize, communicate and mobilize. The nature of the internet itself eliminates the boundaries of the state, challenging the certainty of state sovereignty and control.

Of course, social movements can be crushed by a totalitarian state. This issue is addressed in the section 'Chinese dissidents' (pages 128–143) in Chapter 4, when talking about dissidents using the internet against such states. It is also relevant, for example, in connection to the peace and anti-globalization movements, because 'social movements can easily become segmented, transform themselves into defense of minorities or search for identity, while public life becomes dominated by pro or anti state movements. That is what is happening today, especially in Germany and the US with peace movements' (Waterman 2001: 780).

This section looked at key elements of social movement theory, which can help us with our analysis of sociopolitical cyberconflicts. The subject of new social movements was approached differently by Europeans and Americans. Europeans emphasized the need for a systemic approach that links new forms of social conflict to post-industrial capitalism and the identity-oriented paradigm, while in the US theorists focused on resource-mobilization theory. The key issues in social movement theory analysed here were: the 'newness' of NSMs, the NSMs as a product of a post-industrial economy, the alleged new middle-class orientation of NSMs, the mobilizing structures (the network-style structure of movements using the internet, participation, recruitment, tactics, goals), framing processes (issues, strategy, identity, the effect of the internet on these processes) and the media (and the internet particularly) as a component of the political opportunity structure. This extensive discussion and analysis of central elements of social movement theory is going to prove useful when looking at new social movements and their use of the internet. The next section identifies elements of conflict theory especially relevant to ethnoreligious cyberconflicts.

Conflict theory

> The Other outrages our sense of the kind of nation ours should be in so far as s/he steals our enjoyment – to which we must add that this Other is always an Other in my interior, i.e. that my hatred of the Other is really the hatred of the part (the surplus) of my own enjoyment which I find unbearable and cannot acknowledge, and which therefore transpose ('project') into the Other via a fantasy of the 'Other's enjoyment'.
>
> (Eley and Suny quoted in Salecl 1996: 418–425)

Globalization is often portrayed in broad terms of cultural and technological change: we share friends from different places, culture, food and resources, and we show solidarity with faraway peoples. The problem is that not everybody has access to this opportunity-rich phenomenon. Actually, in certain instances, the idea of globalization is an instance of rhetoric used by governments to justify their submission to financial markets. The same governments are challenged by these global flows of capital, technology, information and people. Also, people might find their identities threatened by such a process. At the same time globalization itself can directly resuscitate local traditions and literally thrive on them.

In other words, the concept of globalization is a conflation of two distinct phenomena: the corporate takeover of the world, and a process of fragmentation of national, ethnic and religious identity-communities, which is creating a more open social context. Whereas the former is to be resisted, the latter provides the basis for a transformative politics, if not a revolutionary macro-politics. The trick is not to try and stop deterritorialization but, rather, to use it as a resource. This has been the practice of global resistance movements, grass-roots social movement organizations and social networks against governments and international institutions. As a result of the activity of such movements, neo-liberal governments and institutions face a counter-hegemonic account of globalization, to which they have responded in a confused and often contradictory way. The challenge expresses itself as resistance and/or violent opposition against governments and against international institutions, either in a sociopolitical or ethnoreligious context. This type of political activity is not sufficiently explained by new social movement theory. Instead, an approach is needed that is capable of capturing the sources and nature of ethnoreligious conflict, as well as its global context. As Moore *et al.* (2005) remind us several liberal theorists have begun to engage with concepts of 'benevolent' empire. If this idea remains unchallenged, it could become normalized within popular culture and eventually become the predominant academic discourse. Therefore, this part of the analysis is influenced by the works of Deleuze and Guattari, Khan, Chesters, Callinicos, Campbell and Jabri, providing another view of global conflict and tools for the analysis of cyberconflicts.

In relation to the ethnoreligious context we should mention here Campbell's definition of ontopology, which provides us with an explanation of complex conflicts like that in Bosnia. Ontopology, as Campbell explains, is the connection of 'the ontological value of present-being to its situation, to the stable and presentable determination of a locality, the topos of territory, native soil, city body in general' (in Edkins *et al.* 1999: 27). Campbell's analysis draws on how Derrida understands conflicts like the one in Bosnia.

In their work *A Thousand Plateaus*, Deleuze and Guattari argue that the information revolution is altering the nature of conflict by strengthening network forms of organization over hierarchical forms. The network form of organization could also be described as the rhizome (1987). They argue against a world where 'the tree is already the image of the world or the root the image of the world tree' (1987: 5). All the various instances of desire, identity and belief are constructed as if they were elements within a single totality, 'arborescent' or 'striated' in Deleuze's terms, like the branches coming from the main trunk of a tree. One of the interesting aspects of their argument is that they prefer to explain the world with principles of connection, heterogeneity and multiplicity, where 'any point of a rhizome can be connected to anything other, and must be' (1987: 7). In this vein, a rhizome establishes connections between semiotic chains, organizations of power and circumstances relevant to the arts, sciences and social struggles. In contrast to centred (even polycentric) systems with hierarchical modes of communication and pre-established paths, the rhizome is an acentred, nonhierarchical, non-signifying system without a General and without an organizing memory or central authority, defined solely by a circu (Deleuze and Guattari 1987: 40).

Based on this rhizomatic conceptualization of the world, the internet is a typical rhizomatic structure and the groups using it are rhizomatic in character because they seem to have no leader, coming together for an event (for example, anti-globalization protests or hacking enemy websites) and dissolving again back to their own ceaselessly changing line of flight into the adventitious underground stems and aerial roots of the rhizome. Deleuze and Guattari might easily have been speaking about the internet when they wrote:

> To these centred systems the authors contrast acentred systems, finite networks of automata in which communication runs from any neighbour to any other, the stems or channels do not pre-exist, and all individuals are interchangeable, defined only by their state at a given moment – such that the local operations are co-ordinated and the final, global result synchronised without a central agency.
>
> (1987: 17)

The world system operates as an arborescent apparatus. However, such an apparatus is necessarily haunted by the possible emergence of 'lines of flight' which take its elements outside the framework it constitutes. The elements which escape the world system have a different structure – less arborescent than rhizomatic, emerging through underground networks connected horizontally and lacking a hierarchic centre. The system's resort to violence is an attempt to crush various rhizomatic and quasi-rhizomatic elements which tend to escape it.

It is no wonder that rhizomes are a source of threat to those whose commitments are structured around the positivist valuation of machines of control. Arguments like these were taken up by Arquilla and Ronfeldt to explore the future of conflict and network forms of social organization. Particularly relevant to my research is the question of how the structure of the internet itself (a global network with no central authority) has offered another experience of governance (no governance), time and space (compression), ideology (freedom of information and access to it), identity (multiplicity) and a fundamental opposition to surveillance and control, boundaries and apparatuses. In the final analysis, new information-age ideologies could easily be arguing for a transfer of virtual social and political structures to the offline world, reversing for once the existing process of online actors imitating real life in cyberspace. The internet is not a medium. It is 'another' place:

> Global interpersonal communication is the greatest tool for world peace our species has ever known. We have the technology to achieve collective consciousness on a planetary scale. The potential of Electronic Revolution is awesome. Instead of electing an aristocracy whose choices are packaged by mass media marketing to govern us, we have the ability to transcend the physical limitations of deceptive appearance and illuminate the truth of being through the digitized reflection of intelligence.
>
> (Taylor 1999)

The form of the internet is itself a message, a symbolic challenge to dominant patterns of hierarchical structures of governance. The wars in Iraq and Afghanistan largely follow the model of non-war, as outlined in Baudrillard's analysis of the first Gulf war and further extended by Ignatieff (2000). Rather than being collisions of two powers located symmetrically within a single discourse, they involve the feints and counterfeints of two sides separated by radical discursive difference. America and its allies were in both cases attempting to impose a discourse of control embodying a logic of deterrence. As General Wesley Clark said of the bombing of Yugoslavia, 'this was not, strictly speaking, a war' (in Ignatieff 2000: 3).

Non-war happens within an almost virtual 'public sphere'. As Baudrillard argues, media images are now the continuation of war by other means. War, the most concentrated form of violence, has become cinematographic and televisual, just like the mechanically produced image.

> The true belligerents are those who thrive on the ideology of the truth of this war, despite the fact that the war itself exerts its ravages on another level, through faking, through hyper reality, the simulacrum, through all these strategies of psychological deterrence that make play with facts and images, with the precession of the virtual over the real, virtual time over real time, and the inexorable confusion between the two.
>
> (Baudrillard 1995: 177)

As Bloom argues, 'the mass public as a foreign policy variable will always react against policies that can be perceived to be a threat to national identity and to policies which protect or enhance national identity ... national chauvinism is commercially successful' (1990: 80). Another interesting point Bloom makes is that there is one tier of information possessed by the security services and decision-making elites, while there is another tier in the public domain. The restriction on complete disclosure is precisely to avoid the possible triggering of the national identity dynamic which would take decision making out of the hands of the 'responsible' and informed few (Bloom 1990: 88).

Non-war often occurs in the form of ethnoreligious conflict, based on issues of ethnic, religious and national affiliation involving a strong sense of identity. Ethnic affiliation provides a sense of security in a divided society, reciprocal help, and protection against neglect of one's interests by strangers. Ethnic divisions reaffirm rather than undermine fixity and closure. As Horowitz puts it, '[i]n deeply divided societies ethnicity – in contrast to other lines of cleavage, such as class or occupation – appear permanent and all encompassing, predetermining who will be granted and denied access to power and resources' (in Diamond and Plattner 1994: xviii). Ali Khan argues that there is an old and apparently primordial human inclination to maintain self-identity by continually creating an 'other'. Power systems become dependent on such identities and loyalties via patronage networks (Khan 1996: 128).

The information revolution is likely to strengthen local and ethnic identities. In fact, territorial and ethnic communities may become even more cohesive and permanent. Communication resources in many nation-states are actually decentralized, creating local and provincial print media, radio and television. Pluralist and alternative viewpoints in national discourse are protected by allocating radio and television resources to ethnic, linguistic and cultural minorities. This resurgence of territorial and ethnic identities may weaken those nation-states that are built by joining disparate historic provinces, localities and ethnic populations (Khan 1996: 128).

Khan's model is certainly relevant to politics in Iraq and Afghanistan, and is echoed in Vivienne Jabri's analysis of the ways in which discourses of inclusion and exclusion constitute individual identity in ways which lead to violent conflicts.

> Since every search for identity includes differentiating oneself from what one is not, identity politics is always a politics of the creation of difference ... What is shocking about these developments is not the inevitable dialectic of identity/difference that they display, but rather the atavistic belief that identities can be maintained and secured only by eliminating difference and otherness. The negotiation of identity/difference to use William Connolly's felicitous phrase is the political problem facing democracies on a global scale.
>
> (Jabri 1996: 3)

Three phenomena challenge these traditional notions of sovereign bodies: corporate transnationalism/global economy, environmental unity and the information revolution are in the process of transforming traditional notions of territorial communities tied to contiguous geographical areas. Information-driven communities require no geographical contiguity, thus threatening the core characteristic of the state structure.

Robin Brown, when considering the relevance of Schattschneider's work to the information society, argues that private matters become identified as public as the scope of conflict grows. Easier access to information, and the mobilizational possibilities that result, make it easier to expand the scope of conflict. As a result, the information society, by facilitating the diffusion of information on a global basis, creates new possibilities for political strategy through the globalization of conflict (Brown 2002: 264–265). As Brown puts it:

> Schattschneider's theory would suggest that groups that are disadvantaged at the current scope of politics would initially embrace the new technology. Dominant actors would seek to limit its impact but if this was not feasible they would have to adopt their own strategies to the new environment.
>
> (2002: 266)

In line with this argument, if some people want to expand the scope of conflict, then others want to reduce it. There are multiple ways to achieve this. 'The first is to ensure that people are unaware of potential issues or actual conflicts. The second strategy is to construct a political discourse that legitimates the involvement of some people but not others' (Brown 2002: 268). An example of this arose in political discourse during the 2003 Iraq war.

Furthermore, between the global resistance movements and the ethno-religious movements, it is clear that very little of the periphery is subsumed into the world system in a stable and hegemonic way; there are, rather, flows of domination between 'rich and poor, hemispherically from N to S, regionally between peripheral and core nationally across class and ethnic boundaries' (Chesters: 42–65). The 'line of flight' emerging from resistance movements could be best described as parallel lines, a parallelogram of forces as Graeme Chesters argues: the whole of singularities against WTO, IMF, WB; People's Global Action, World Social Forum and such like have an ecology of action which indicates a web of horizontal social solidarities in which power might be devolved, or even dissolved. Anti-capitalist and other rhizomatic groups have constructed many new forms of political action, and also new forms of communication.

In theory that is. In practice when we abstract the ethnoreligious movements or groups, we are left with anti-war groups, anti-capitalist/anti-globalization movements, campaigns that focus on specific issues and grievances, and lifestyle-issue movements (the latter not quite relevant to this discussion, as pointed out by Callinicos (www.swp.org.uk/INTER/regroupen.pdf). These can be referred to as movements of a sociopolitical nature. In contrast to the closure of space, the violence and identity divide found in enthnoreligious discourses, these movements seem to rely more on networking and grass-roots organizing, much more than is the case with the hierarchical structures, states and their followers. Several metaphors have been used to describe a large number of groups being brought together under a common cause, groups that disperse as easily as they come together, a parallelogram of forces following a swarm logic, like ants in an ant colony. It may be argued that because citizens do not believe in power through conventional politics, citizens are increasingly sympathetic to direct action.

Without the need for a leader, and without a particular individual who either has a privileged insight or is able to conceptualize the characteristics of the whole, there is an emphasis on participation, antipathy to hierarchy, a preference for consensus processes and/or directly democratic decision-making, an ethos of respect for differences and an assertion of unity in diversity. The project is less the capture of the state apparatus than the construction of an open and transnational public sphere, a rhizomatic extension of struggles operating through weak ties (Chesters: 42–65).

Sometimes, ethnoreligious and sociopolitical conflicts overlap. Whatever the connection between the Palestinians' oppression and global capitalism in the shape of US imperialism, the system itself is not the centre of Palestinian consciousness when they fight the Israeli state. As Trotsky pointed out, sometimes the point of honour is not what one has in common with the movement, but in the particularity that distinguishes them from it (www.marxists.org/archive/trotsky/1923-nc/index.htm). Apart from the rhizo-terrorists such as al-Qaeda, anti-capitalist and other rhizomatic groups have constructed many new forms of political action, and also new forms of communication.

Furthermore, another part of conflict theory deals with conflict resolution. I think it will prove useful for a later discussion in the section on pages 167–170 in Chapter 5 to deal here with some aspects of conflict resolution, such as mediation and negotiation.

To begin with, Tidwell (1998: 156) refers to the work of Bush and Folger, *The Promise of Mediation*. The authors argue that there are four primary objectives found in mediation, broadly defined. The first is satisfaction theory, wherein mediation serves to satisfy human needs. The second is social justice, which emphasizes the role of mediation in the formation of community. The third perspective is transformation, where mediation is the ability to transform both individuals and society as a whole. Transformation occurs when people alter their values and beliefs about themselves and others. Lastly, there is the oppression story, which views mediation as a tool for control and domination.

Another way of problematizing negotiation is offered by what Rothstein calls the concept of interactive problem-solving (1999: 193). According to him, negotiation should first of all treat the conflict as a problem shared by the parties involved. Second, negotiation explores ways of solving this problem, 'not by eliminating all conflict and potential conflict between the parties, but by addressing the underlying causes of the conflict and reversing the escalatory dynamics of conflict relationship'. His third step is that negotiation is an interactive process, capable of producing problem-solving ideas which respond to the parties' fundamental concerns.

Furthermore, negotiation is very much affected by culture. Culture is brought into the negotiation process by individuals, groups and organizations. It conditions how they view the negotiation, the kind of game they perceive to be going on. As Berton and his colleagues argue, structural components of a negotiation are not culturally free (Berton 1999: 21, 26). External constraints, such as the legal framework and the organizational setting of a negotiation, are social constructs. The other main factors affecting negotiation include the number of parties involved, the number of issues at stake and the distribution of power between the parties. Also, beliefs express a set of values derived from the cultural background of the negotiator. In this way, cultural values directly affect the behaviour of actors involved. Tidwell has the same view when he argues that 'methods such as mediation or facilitation may not be appropriate within a given cultural context ... [I]n cultures in which to speak directly about a conflict is regarded inappropriate, many Western methods would simply not work' (1998: 6).

According to Cairns, there is no single way to build a peaceful society after a war has ended. What both 'successful' and 'failed' peace processes show is that 'peace cannot be built simply on multi-party elections, after which most of the international community make a quick exit' (Cairns 1997: 88). What he recommends instead is reviving the community systems of mediation. This can be an effective means of resolving current conflicts. He uses the example of community leaders from both sides of Mali's conflict to make his point. They have formed so-called peace cells to resolve intercommunal disputes. This initiative has not replaced traditional ways of dealing with conflicts, but it is built on them, including the web of alliances, which cross ethnic groups and families, known as 'cousinage'. Another positive case has emerged in Colombia, with the growth of so-called 'communities of peace'. These groups declare neutrality in the fighting between military, paramilitary and guerrillas. The Antioqua Indigenous Organization, which Oxfam supports, combines practical help of food shelter and medicine with the use of the media to proclaim neutrality.

This section discussed the rhizomatic structure of new social movements and of the internet itself in the contexts of globalization, issues of national identity, ontopology, ethnic affiliation and discourses of inclusion and exclusion. These issues are crucially linked to ethnoreligious cyberconflicts and the various ways the opposing parties develop and perform. This is followed by an examination of the structure of conflicts involving new social movements such as the anti-war and anti-capitalist movements. These movements feature an emphasis on participation, antipathy to hierarchy, and a focus that lies less in the capture of state apparatus and more in the construction of an open and transnational public sphere. This discussion provided a context through which to analyse new social movements and their use of the internet in sociopolitical cyberconflicts, and a focus on national identity and discourse to help interpret ethnoreligious cyberconflicts. Lastly, some of the aspects of negotiation and mediation as part of conflict resolution techniques will be discussed, which will prove useful in the study of attempts at conflict resolution on the internet. The next section offers a discussion of media theory elements that should prove useful when looking at the internet as a medium.

Media theory

'The Gulf War is over and the press lost.'[1]

The total obliteration of war by information, propaganda, commentaries, with camera-men in the first tanks and war reporters dying heroic deaths, the mish-mash of enlightened manipulation of public opinion and oblivious activity: all this is another expression for the withering of experience, the vacuum between men and their fate, in which their real fate lies. It is as if the reified, hardened plaster cast of events takes place of events themselves. Men are reduced to walk-on parts in a monster documentary film which has no spectators, since the least of them has his bit to do on the screen.

(Poster 1995: 55, quoting Adorno 1974)

De Fleur Ball-Rokeach (1982) distinguishes between three different effects of the mass media on individuals, which are closely linked to the emergence and coverage of global conflicts. These include cognitive effects (the creation and resolution of ambiguity, attitude formation, agenda-setting, expansion of people's systems of beliefs, impact on values); affective effects (desensitization, anxiety, morale and alienation); and behavioural effects (activation, deactivation). Another particularly relevant issue is the dynamic of mass media construction of consensus, control, adaptation, conflict and change. These dimensions and their effects on conflict of either ethnoreligious or sociopolitical roots are the focus of this section.

One of the key issues is the media's method of setting an agenda. A useful approach looks at the actual language used by the media, especially appropriate for internet analysis because the most common feature of the internet is text. The questions that need to be asked, as outlined by Fairclough (1995: 5), are as follows: How is the world (events, relationships) represented? What identities are set up between those involved in the story? What relationships are set up between those involved? What, then, are the particular representations of the world, particular constructions of social identities and particular constructions of social relations? Among others, Fairclough provides an example of the analysis of discourses in texts when he examines press coverage of an air attack on Iraq by the USA, Britain and France on 13 January 1993, referring to five British newspapers. A brief glimpse of this analysis could prove useful as a guideline for this research:

The main headline and lead paragraph from the *Sun* show that formulations of the attack do not by any means draw only upon military discourse: Spank you And Good Night … and More than 100 allied jets … gave tyrant Saddam Hussein a spanking. This is a metaphorical application of an authoritarian discourse of family discipline which is a prominent element in the representations of the attack – Saddam as the naughty child punished by his exasperated parents.

(Fairclough 1995: 95)

The Gulf war has inspired a lot of media research, being the ultimate television war (Taylor 1992; McGregor 1997; Wolfsfeld 1997) and it will continue to do so in the future, because it was the first war extensively covered through the internet. Taylor poses key questions in relation to the control of information by military and political authorities: What arrangements were made for the release of information and why? How much censorship was taking place? How far back did the journalists stand from what was being told to them or were they merely drawn into the media management system? What alternative sources of information were available? (Taylor 1992: vii). This last question is especially interesting in relation to web coverage and blogging during the 2003 war in Iraq.

A good example of the media construction of social identities, and specifically the enemy identity, is the emphasis placed by Bush senior's administration on projecting the image of an enemy posing a serious military, economic and ideological threat to the new world order. Saddam was represented as a formidable military power, ready to dominate the Middle East. Taylor puts it very graphically: 'Kuwait had been "raped", Iraqi troops were "plundering" the tiny and helpless state and "butchering" its people ... It was even claimed for domestic consumption, that this would be a war for the American way of life. Saddam must not be "appeased" as Hitler had been in Munich ...' (1992: 5)

The fact that CNN provided a public insight into the traditionally secretive world of diplomacy has altered the way modern warfare is projected onto the world's television screens. What is more, it raises a lot of ambiguity over what truly happened in that war – that is, over the relationship between the 'real' war and the war as portrayed by the media.

One of the central questions is how censored or restricted journalists are while covering conflicts and how 'unbiased' they can be in constructing social reality, identities and relations. Knightley, in his work on war correspondents from Crimea to Kosovo, reveals how the role of the war correspondent as the heroic truth-seeker is in danger of becoming more that of a myth-maker. Again, the Gulf war conflict is taken as a turning point in the history of war correspondents. As Knightley argues, '[n]ot only was it a war in which the military succeeded in changing people's perceptions of what battle was really like, but one in which the way the war was communicated was as important as the conduct of the war itself' (2000: 500).

War correspondents are just one part of a larger picture. As McGregor (1997) argues, the largely consistent findings show the news products to be, in one sense or another, an artificial and very predictable symbolic construction of reality. However, this conclusion is itself open to alternative interpretations, since homogeneity could result either from a hegemonic ideology or simply from the standardization to be expected in mass production processes, or perhaps a combination of the two (McGregor 1997: 78, citing McQuail). In the 'first living room war', the Vietnam war, despite the fact that the network news deserve credit for the eventual disillusionment with the war, at the same time they were also responsible for creating, or at least reinforcing, the illusion of American omnipotence in the first place (Epstein quoted in Mercer *et al.* 1987:

229). An example of this was that American media delayed for two years the reporting of the My Lai massacre, not because of censorship, nor because the facts were not instantly available, but due to resistance to the story by the US media itself (Mercer *et al.* 1987). That was because the massacre occurred in 1967, when the storyline was focused on 'good news' about a war which editors were persuaded the US was winning:

> Other television reporters have detected significant media impact 'only at moments of policy panic' or 'where policymaking is weak or cynical'. Although television assumed a high profile in accounts of policy influence, closer analysis revealed interdependence with other forms of news media.
>
> (Koppel Congressional Hearing 5)

A much more recent 'construction of reality' by Western media was in the Balkan conflicts, where civil wars were called 'ethnic cleansing' and painted with simple terms of 'goodies and baddies', portraying the situation as a conflict where the international community could do nothing. When Milosevic failed to 'stop the ethnic cleansing' of Albanians in Kosovo and the decision to bomb him into submission was taken, an impressive system of control and propaganda in both the US and Britain swung into action (Knightley 2000: 502). In the two main sections of interest, news of the fighting and justification for it, the Kosovo conflict was a case in point of propaganda-led warfare, because 'to sell a war in a democracy when you are not attacked, you have to demonise the leader or show that there are humanitarian reasons for going in' (Lichter quoted in Knightley 2000: 502). The extent to which the media were managed was high, as the words of Alastair Campbell, press officer and adviser to Blair, betray: 'it was vital to try and hold the public's interest on our terms' (in Knightley 2000: 503). The same Mr Campbell was later accused by the media of spindoctoring, through what came to be known as the 'dodgy' dossier, a report which relied in part on a PhD thesis on the web, justifying Britain's decision to go to war in Iraq in March 2003, leading to two inquiries that involved the Prime Minister himself (Deans 14 July 2004). Tony Blair has also claimed a top prize, winning the 'Lifetime Menace' award for what the London-based Privacy International characterized as 'his active involvement in the government's attack on civil liberties', as he angered privacy groups with his plans to force phone companies and internet service providers to retain users' data for twelve months as part of the state's stepped-up war on terrorism and crime (Scheeres 25 March 2003).

The implementation of policy decisions is often affected by the weight of media exposure. As Major General Lewis MacKenzie, commander of UNPRO-FOR in Sarajevo, commented:

> Wherever the media goes, a lot of serious violations of human rights either move away or stop. The media was the only major weapon system I had. Whenever I went to negotiations with the warring parties, it was a tremendous weapon to be able to say: 'OK, if you don't want to do it the UN's

way, I'll nail your butt on CNN in about 20 minutes'. That worked, nine times out of ten.

(*New York Times* 18 July 1995 quoted in Minear *et al*. 1995: 59)

Furthermore, there are three levels of media effect on policy: strategic (decisions on whether to intervene, withdraw, etc), tactical (innovations in the protection of Sarajevo that responded in part to media coverage) and presentational (they probably would not have happened apart from the media exposure that would accompany them). In addition, there are different degrees of media effect: primary (Somalia, Haiti, Rwanda), secondary (contributory effect, e.g. Iraq and Bosnia), and negligible (Minear *et al*. 1994: 71–72).

Moreover, as far as humanitarian intervention is concerned, the media had been criticized on several fronts. One complaint is that they pay too much attention to breaking events, while ignoring the historical and political context (reportage of the Rwandan crisis). Another is that the international media have focused on subjects of perceived interest to readers and viewers in developed countries, denigrating local institutions and overemphasizing the importance of international and western initiatives (Minear *et al*. 1994: 37). More importantly, criticism often centres on the tendency of the media to perpetuate negative images and ethnocentric views, when those who suffer are often non-white and their 'rescuers' white, a portrayal which contributes to charges of racism in war coverage, or to media coverage as the 'pornography of suffering'.

In a media-intense environment, politicians and the public have become very unforgiving of even minor mistakes and transgressions. Events with minor operational effects, such as the killing of a Somali youth for stealing a soldier's sunglasses or the dramatic rescue of a downed F-16 pilot in Bosnia, often have disproportionately large effects on public opinion and therefore policy and outcomes. As a result, even the minutest aspect of military operations must now be planned with sensitivity to the public perception of the fight (Shapiro in Khalizad *et al*. 1999: 125).

Nationalism, constructed primarily (as Anderson shows) through traditional media such as newspapers and novels and also promoted by newer media such as TV, causes the entire populace, not just the elite, to identify its interests with those of the government. Consequently, the government can mobilize the entire capacity of a society for a prolonged war; and even if the regular army is defeated, the people will continue to resist through irregular means. The inability of the government to control the information flow gives enemies a means to undermine this identity of interests. New techniques that allow the manipulation of video images and sound recordings and therefore allow the conduct of sophisticated psychological operations provide another resource for undermining the identity of interests between the government and the wider populace. Indeed, some believe that the real war in the information age will be for the hearts and minds of the populace or the fears and insecurities of the troops (Shapiro *et al*. 1999: 126).

But information technology is linked to the inventiveness, freedom, aspirations, and irrepressibility of the citizen. If anything, state power, in its traditional sense, can only retard this technology. The information revolution both liberates and requires liberation. As the US experience shows, the freer the market, the greater the level of performance that information technology delivers (Gombert 2001: 50).

An extreme example of the possibilities and impact of the individual in virtual political reality occurred during the Chechen hostage crisis in the Moscow theatre in October 2002, when one of the hostages, Rankov, contacted his friend, Olga Brukovsky, on a mobile phone while the stand-off was in progress. She took down his words and published them online at LiveJournal.com, a website that is popular amongst Russians, initiating a flurry of responses especially when Russian security police raided the theatre in a controversial rescue attempt, killing 117 hostages. Anton Nossik, one of the founders of LiveJournal's Russian community and chief editor of Lenta.Ru, the country's leading online news service, said at the time that the site had become an especially important source of information for people living in remote locations, such as Siberia and the Far East, where news agencies seldom send reporters. While Russian authorities attempted to control the flow of information (one television station was temporarily shut down, along with the website for the radio station Echo of Moscow), LiveJournal served as 'a good mirror of public opinion' according to Roman Leibov, a professor of Russian literature at Tartu University (in Kettman 18 July 2001).

In relation to the LiveJournal experience, the Europeans after having watched a number of heavily hyped US internet publications struggle have chosen more modest and long-term business plans. For example, Transitions Online, based in Prague, after it ran out of money as a print publication, sought refuge on the internet. The publication, which charges its subscribers a small fee but still raises 90 per cent of its budgets through grants, has a network of local correspondents in more than two dozen Central and Eastern European countries and has won a NetMedia award in 2001 for outstanding contribution to online journalism in Europe. The vast majority of their correspondents had never met until a conference was organized (Kettman 18 July 2001).

The online records of activity are commonly falsified either by direct modification of the records themselves or by replacement of the monitoring software that produces these records. While there are analogies to these activities in the physical world, the ease, rate and invisibility of these activities on the internet especially complicates the analysis task. Direct support for dissidents or embryonic democratic institutions is increasingly available both from the governments and non-governmental organizations of the democratic core. The penetrability of even self-isolated societies is growing, especially when sophisticated transnational 'civil society' groups make it their business to network with the oppressed.

Advanced technological systems will not only help shape the environment of future conflict but will also magnify the importance of the psychological battle to conflict outcomes. Leaders of the several former Yugoslav republics used television, radio and print media to promote ethnic hatred and mobilize their publics to take up arms to advance or defend communal political and territorial interests. Indeed, some observers believe the media became the 'main instruments in stirring up and managing' the conflict in the former Yugoslavia. Similarly, broadcasts from the government-controlled Rwanda Radio did much to foster the 1994 genocide in Rwanda by deliberately fomenting ethnic hatred among the Hutus and inciting the mass killings of Tutsis. After the Hutu government had been routed by Tutsi forces, a mobile radio still under the control of former Hutu government officials precipitated the massive flight of Hutu refugees into Tanzania and Zaïre by assuring them that they faced 'certain slaughter' if they fell under Tutsi control (Adelman and Suhrke 1996: 38). Hostile radio broadcasts also helped to undermine the US and UN intervention in Somalia. To counter US and UN attempts to marginalize him politically, Aideed successfully used his radio station in Mogadishu to rally support for his continued leadership and to foment anti-US and anti-UN sentiment among his countrymen (Hirsch and Oakley 1995: 116–117).

The US and other news media will become an increasingly ubiquitous presence on the future battlefield. The media will have an independent capability to gain access to future conflict arenas and to provide real-time visual and audio coverage of battlefield events. Thus, the media will be able to report promptly the human costs of US combat involvement to both US domestic and international audiences. Evidently, embedded journalists with the troops during the 2003 Iraq conflict is a step in that direction. As the US experience in Vietnam and Somalia demonstrated, media news coverage and commentary will help shape US domestic perceptions about whether a US military involvement is effective or not and, most importantly, whether it merits continued public support (Hosmer in Khalizad *et al.* 1999).

Advanced technological information systems will allow state and substate actors, including news services, non-governmental organizations, and even individual citizens, to make voice, video and written information instantly available to audiences located in the remotest areas of the globe. 'ICTs have sometimes played a dramatic role in communicating rich, multiple impressions of particular events as they unfold . . ., but they have been used less successfully in promoting a coherent, collective assessment of what these events mean within the overall process of social change' (Wright 2004: 90).

The idea of this thesis is to try and look at media and conflict together and focus on their interaction, followed up by more analysis on ethnoreligious conflict as a phenomenon and how it is affected by the presence of the media. With the internet ranking as the top information source, outpacing TV, newspapers and radio – as found by a UCLA study – media theory should relate, analyse and discuss the new medium in a more rigorous manner.

A theoretical model that attempts to discuss media and conflict together is Gadi Wolfsfeld's political contest model (1997: 3–5). The thrust of this model is that the best way to understand the role of the news media in politics is to view the competition over the news media as part of a larger and more significant contest among political antagonists for political control. Wolfsfeld's model rests on five major arguments. First, that the political process is more likely to have an influence on the news media than the news media on the political process. Second, the authorities' level of control over the political environment is one of the key variables that determine the role of the news media in political conflicts. Political conflicts are characterized by moves and counter-moves, as each antagonist tries to initiate and control political events, to dominate political discourse about the conflict, and to mobilize as many supporters as possible to their side. Those who have success in these areas also enjoy a good deal of success in the news media. Third, the role of the news media in political conflicts varies over time and across different circumstances. It varies along with such factors as the political context of the conflict, the resources, skills and political power of the players involved, the relationship between the press and each antagonist, the state of public opinion, the ability of the journalists to gain access to the conflict events and, lastly, by what is happening in the field. The fourth argument is that those who hope to understand variations in the role of the news media must look at the competition among antagonists along two dimensions, one structural and the other cultural. The structural dimension looks at the extent of mutual dependence between the antagonists and each news medium to explain the power of each side in the transaction, while the cultural dimension focuses on the construction of media frames of conflict events. Wolfsfeld's fifth argument is that while authorities have tremendous advantages over challengers in the quantity and quality of media coverage they receive, many challengers can overcome these obstacles and use the news media as a tool for political influence.

More analytically, according to Wolfsfeld, the ability of an antagonist to control the political environment can be understood in terms of three variables: the ability to initiate and control events, the ability to regulate the flow of information and the ability to mobilize elite support (Wolfsfeld 1997: 25–28). The first factor relies on the fact that governments are in a much better position to coordinate their press relations when they can anticipate the events that will be covered. When, on the other hand, the powerful are forced to react to events, it suggests that others are setting and framing the media's agenda. The second variable is the ability to regulate the flow of information. Governments, both democratic and non-democratic, often find compelling reasons to employ censorship during political conflicts and this increases the value of official sources of information by eliminating competition. The ability of the powerful to regulate the flow of information to the press is also affected by the nature of the logistical and geographical environment. Powerful governments prefer to operate under conditions in which they can isolate the areas of actual conflict and regulate the entry and exit of journalists. While the physical circumstances of certain locales facilitate government control, other locations are more porous and offer easier access for reporters, thereby increasing the level of journalistic independence. The third variable that decides the powerful's level of control over the information environment is the ability to mobilize elite support:

When the various factions within a government are promoting different frames about a conflict, it is more difficult to control the informational environment because journalists can choose among a variety of sources. When on the other hand, the official frame is the only frame available among elites, journalists will have little choice but to adopt the frame.

(Wolfsfeld 1997: 29)

The construction of media frames is another important issue in Wolfsfeld's work. He claims there are three major elements that contribute to the construction of media frames of conflict: the nature of the information and events that are being processed; the need to create a good news story; and the need to create a story that resonates politically within a particular culture.

The news media have a large variety of frames waiting on the shelf for those activists who are skilled enough to construct an effective package and lucky enough to be promoting them at a time when the authorities are vulnerable to attack. In these cases the news media can play a critical role by legitimating oppositional frames that increase the status, resources, and power of challengers.

(Wolfsfeld 1997: 55)

The 1991 Gulf conflict represents an excellent example of a case in which the news media enthusiastically adopted the authorities' law and order frame and virtually ignored the injustice and defiance frame being promoted by the challenger. The Allied forces enjoyed a multitude of advantages in the structural field that allowed them to control a great deal of the media discourse about the war. Saddam Hussein was defined early by the news media as the aggressor and the international consensus around the law and order frame made it extremely difficult for Iraq to promote its particular version of the injustice and defiance frame (Wolfsfeld 1997: 192).

This section has looked at various elements of media theory such as media discourse and construction of social identity and relations, citing the first Gulf war; the construction of reality by Western media in the Balkan conflicts; the relationship between the media and humanitarian intervention; and the effect of the internet on conflict (the LiveJournal experience). The next section combines elements of the theories analysed so far, deriving an analytical framework for cyberconflict.

Deriving an integrated framework for cyberconflict

This part is bringing together the three theories – social movement, conflict and media theories – into an integrated model for analysing cyberconflict in its ethnoreligious and sociopolitical dimensions. First, I provide a justification of why these theories are necessary for any explanation of cyberconflict. Second, I lay out a proposed model for such an analysis drawing on these theories. Finally,

I call for a reversal of the two modalities of cyberconflict, from hierarchical to rhizomatic and (almost) vice versa.

Reasoning

The central argument of this book is that there are two types of political conflict on the internet (cyberconflict): ethnoreligious and sociopolitical. In order to analyse 'real' political conflicts, political scientists have devised 'conflict theory' and 'international conflict analysis'. The problem of just using conflict theory is that it cannot fully account for two important parameters of cyberconflict: its sociopolitical dimension (social movements or dissidents using the internet against antagonistic institutions) and the fact that conflict is taking place within or with the help of a medium (the internet). Again, in order to understand how social movements are engaging in cyberconflicts and are affected by the use of IT technology, it would be logical to use the theoretical tools already put in place by new social movement theorists. The problem with this approach is that it leaves out a more thorough examination of the media context (in this case cyberspace) in which sociopolitical cyberconflicts are taking place. Again, if one decides to just use media theory to analyse cyberconflict, many questions about what the conflict means for the ethnoreligious and sociopolitical groups involved will remain unanswered. Lastly, an attempt to integrate social movement and media theories, while leaving conflict theory out, will miss explanations of ethnoreligious conflict. The reason for this is that social movement theory, as far as it was discussed and analysed above, does not seem to engage with conflict in ethnic and religious terms. In other words, any attempt to analyse an ethnoreligious conflict (such as the Israeli–Palestinian conflict) using mobilization theory simply seems implausible.

The political environment of the internet is analysed here, not in terms of the internet as a mass medium in the traditional sense (i.e. what would be an internet news bureau's or the online version of an already established medium's influence on the political outcome of a conflict), but rather as a significant new resource used by the opposing parties in a conflict. This presents us with a major theoretical challenge, because there is no theoretical model to date that can provide us with the conceptual tools to analyse the use of the internet by the actual parties in the conflict (endogenously) and not just theorize about the way in which the media influence – or do not influence – the political outcome of a conflict (exogenously). The reason for this, as we saw in the section 'A postmodern medium?' (pages 39–50) in Chapter 1, is that the internet is not a traditional medium which groups, institutions or states compete to access. It may be used by anyone, at any time, from most places on the planet. Furthermore, despite the fact that actors using the internet might still seek traditional political goals like power, participation or democracy, the postmodern nature of the medium makes a more complex theoretical approach necessary. Explaining cyberconflict in a single framework, for instance a media studies approach, a conflict theory approach or even a social movement approach such as resource mobilization theory (RMT) would provide us with a one-dimensional and inadequate discussion. My interest lies in examining this phenomenon under as many 'theoretical lights' as possible and at the same time refraining from being blinded by the radiance of the novelty of the medium itself.

Theoretical framework

Search for a satisfactory theoretical framework, has yielded the following parameters to be looked at while analysing cyberconflicts:

1   *Environment of conflict and conflict mapping (real and virtual).* The world system generates an arborescent apparatus, which is haunted by lines of flight, emerging through underground networks connected horizontally and lacking a hierarchical centre (Deleuze and Guattari). The structure of the internet is ideal for network groups (since it is a global network with no central authority) and has offered another experience of governance (no governance), time and space (compression), ideology (freedom of information and access to it), identity (multiplicity) and fundamentally, an opposition to surveillance and control, boundaries and apparatuses. However, in ethnoreligious cyberconflicts, where the groups' systems of belief and organization aspire to hierarchical apparatuses (nation, religion, identification with parties and leaders), this network form is not always evident. This is why there is a dual modality of cyberconflict: one rhizomatic and one hierarchical.

2   *Sociopolitical cyberconflicts.* The impact of ICTs on: a) mobilizing structures (network style of movements using the internet, participation, recruitment, tactics, goals); b) framing processes (issues, strategy, identity, the effect of the internet on these processes); c) political opportunity structure (the internet as a component of this structure); d) hacktivism.

3   *Ethnoreligious cyberconflicts:* a) ethnic/religious affiliation, chauvinism, national identity; b) discourses of inclusion and exclusion; c) information warfare, the use of the internet as a weapon, propaganda and mobilizational resource; d) conflict resolution, which depends on the legal and organizational framework, the number of parties and issues, the distribution of power, and the content of values and beliefs.

4   *The internet as a medium:* a) analysing discourses (representations of the world, constructions of social identities and social relations); b) control of information, level of censorship, alternative sources; c) political contest model among antagonists – the ability to initiate and control events, dominate political discourse, mobilize supporters (Wolfsfeld); d) media effects on policy (strategic, tactical, representational).

Calling for a reversal: from hierarchical to rhizomatic and (almost) vice versa

In their work *Multitude* (2004) Hardt and Negri explain that, in the latter part of the twentieth century, protest movements and revolts followed two primary models. The more traditional one is based on the identity of the struggle, and its unity is organized under central leadership, while the second model is based on the right of each group to express its difference and conduct its own struggle autonomously. Thus, we are either united under the central identity or separate struggles that affirm our differences. The multitude that Hardi and Negri refer to 'replaces the contradictory couple identity-difference with the complementary couple commonality-singularity' (2004: 217; see also the sections 'Social movement theory' (pages 153–171) in Chapter 2 and 'Sociopolitical cyberconflicts' (pages 121–127) in Chapter 4 on the divergence of groups and ideologies comprising the anti-globalization movement). Nevertheless, the mobilization of the commons has taken divergent organizational forms. Despite the network forms of various contemporary global uprisings, there are very important differences:

> The new global cycle of struggles is a mobilization of the common that takes the form of an open, distributed network, in which no center exerts control and all nodes express themselves freely. Al-Qaeda, experts say, is also a network but a network with strict hierarchy and a central figure of command. Finally, the goals are too diametrically opposed. Al-Qaeda attacks the global political body in order to resuscitate older regional social and political bodies under the control of religious authority, whereas the globalization struggles challenge the global political body in order to create a freer, more democratic global world.
>
> (Hardt and Negri, 2004: 218)

What the reversal argument identifies further is a dual modality: if ethnoreligious cyberconflicts are mapped as representing/defending loyalties of hierarchical apparatuses and sociopolitical cyberconflicts are empowering network forms of organization, then the following results. Actors in ethnoreligious CC need to operate in a more *networked/multitudinal* fashion, if they are fighting network forms of terrorism or resistance, or if they are clandestine networks attempting to influence the global political environment. Actors in sociopolitical CC need to operate in a more *organized/conscious* fashion, if they are to constructively engage with the present global political system or parts of that system. Conflict resolution will only be possible when hierarchical apparatuses become more networked and rhizomatic groups become more conscious of the rest of their hosting network.

Following Arquilla and Ronfeldt (2001) a further observation that can be made here is that ethnoreligious dimensions to cyberconflict environments are going to prove crucial to future 'high' information warfare, targeting national and international infrastructures. Sociopolitical dimensions in both real and virtual environments are equally relevant to future 'low' societal end of the spectrum warfare, targeting real international bodies and movements and their virtual representatives.

Sociopolitical cyberconflicts and the argument for greater organizational efficiency

The majority of groups engaging in sociopolitical cyberconflicts, as we see in the discussion of the anti-globalization/anti-capitalist movement in the section 'Sociopolitical cyberconflicts' (pages 121–127) in Chapter 4 and in the study of the anti-war movement in the section on pages 175–178 in Chapter 6, can be placed under the category of transnational social movements (TMs). Movements are changing from fairly coherent national organizations into transnational networks, with highly fragmented and specialized nodes composed of organizations and less organized mobilizations, all of which are linked through new technologies of communication (Tarrow 1998: 178, citing Garner 1994: 431).

Tarrow defines TMs as 'sustained interactions with opponents – national or nonnational – by connected networks of challengers organized across national boundaries' (1998: 84). In these movements we find what is called 'cross-border diffusion': the communication of movement ideas, forms of organization, or challenges to similar targets from one center of contention to another (1998: 86). Tarrow's preoccupation with transnational movements highlights the difficulties of aggregating people with different demands and in different locations in concerted campaigns of collective action. This involves 'mounting collective challenges, drawing on social networks, common purposes, and cultural frameworks, building solidarity through connective structures and collective identities to sustain collective action' (1998: 4). The dangers resulting from such an attempt are, first, the search for transnational common denominators that will resonate at some level with many cultures and traditions, and second, following a variety of issues that take root in particular places, which can produce ideological divergence within the same transnational network, as activists adapt the network to their cultures (1998: 191).

These dangers are indeed present for transnational movements. The aim here is to demonstrate the need to strengthen the organizational structures and informal, weak connective structures which these non-hierarchical movements normally have. Without sufficient organization – whether formal or informal – political opportunities are not likely to be seized. An emerging international pattern of social movement organization seems to be appearing: a combination of small professional leaderships; large but passive mass support; and impersonal network-like structures (Tarrow 1998: 133).

One of the explanations of why new social movements are decentralized or rhizomatic is that strong executive power structures in a given political system tend to induce a fundamental critique of bureaucratic and hierarchical political forms, which are then reflected in the movements' emphasis on informal and decentralized structures (Rucht 1996: 192). Explaining this, however, does not imply that the following can be ignored: 'SMOs [social movement organizations] with formalized and professionalized structures tend to have easier access to public authorities, because government bureaucracies prefer to deal with organizations with working procedures similar to their own' (Kriesi 1996: 158).

As it happens, Bennett argues that new waves of movements (with their variform and shifting organizations, their tendency to produce rapid and rapidly liquidated coalitions, and their focus on short- and medium-term issues rather than fully fledged ideologies) do not produce standing activist commitments or deeply held loyalties (2004: 128). Although this sounds largely convincing, one reservation about Bennett's argument should be considered. Rapid coalitions can also have a positive impact on the flexibility and mobility of these movements, particularly in the case of the anti-globalization movement. Such coalitions have greater discretionary resources, enjoy easier access to the media, have cheaper and faster geographic mobility and cultural interaction, and can call upon the collaboration of different types of movement-like organizations for rapidly organized issue campaigns (Tarrow 1998: 207).

Brown, following Tarrow, has expressed the hierarchical/network dilemma in similar terms (1999: 9). His argument is that networks

> with decentralized decision-making have the advantage of being better able to deal with local conditions than centralized and hierarchical organizations. However, networks may find it difficult to develop cultures or perspectives and while their superiority to hierarchies is asserted the advantage of hierarchy is that it allows concentration of resources behind a common purpose quickly and easily.
>
> (2000: 11)

Following this reasoning, the most illuminating perspective is provided by Tarrow, who explains that the problem for movement organizers is to create organizational models that are sufficiently robust to structure sustained relations with opponents, but are flexible enough to permit the informal connections that link people and networks to one another to aggregate and coordinate contention. The most effective forms of organization are based on partly autonomous and contextually rooted local units linked by connective structures, and coordinated by formal organizations (Tarrow 1998: 124). He develops the argument further:

> The dilemma of hierarchical movement organizations is that, when they permanently internalize their base, they lose their capacity for disruption, but when they move in the opposite direction, they lack the infrastructure to maintain a sustained interaction with allies, authorities, and supporters. This suggests a delicate balance between formal organization and autonomy – one that can only be bridged by strong, informal, and nonhierarchical connective structures.
>
> (1998: 137)

One may then plausibly argue for stronger organization with strong informal and non-hierarchical connective structures which, although remaining autonomous, would be coordinated by formal organizations. This might solve some of the problems stated above. It is equally important to stress the impact of new communication technologies on these movements, because in several instances the internet has been responsible for the rapid cross-border diffusion of movement ideas, the organization of protest and even the globality of protest itself:

> The growth of broad networks despite (or because of) relatively weak social identity and ideology ties, the transformation of both individual member organizations and the growth of patterns of whole networks, and the capacity to communicate messages from desktops to television screens. The same qualities that make these communication-based politics durable also make then vulnerable to problems of control, decision making and collective identity.
>
> (Bennett 2004: 144)

This is why stronger organization is of paramount importance. In order for the communication revolution to change the political opportunity structure and challenge the scope of political life, 'rather than assuming that the impact of technology makes organization irrelevant, this suggests that the fragmentation of media spheres will actually make organization more important as a way of triggering political action' (Brown 2002: 271).

As Van Aelst and Walgrave have argued, although the availability of new communication technologies makes traditional organizations somewhat dispensable for mobilization purposes, a certain institutionalization remains necessary in order to exert a more lasting political influence (Edwards 2004: 200). But if the network is the prevalent organizational form surfacing from the integration of the internet, the movement's goals can be achieved only when combined strategies of traditional and new media usage are implemented (Van de Donk *et al.* 2004: 161). The reason for this is that 'while networked communication may help sustain the campaigns that organize global activism, these leaderless networks may undermine the thematic coherence of the ideas that are communicated through them' (Bennett 2004: 134). Moreover, communication in diverse networks is ideologically thin and 'as anyone who has caught the internet virus can attest, virtual activism may serve as a substitute – and not as a spur – to activism in the real world' (Tarrow 1998: 193).

Before concluding that the world is entering an unprecedented age of global movements, we will 'need to follow some of the recent campaigns that have been assisted by electronic communication to find out whether it increases the movement's power or merely changes how it frames its message' (Tarrow 1998: 194). Three kinds of long-term and indirect effects of movements are important: their effect on the political socialization of the people and groups who have participated in them; the effects of their struggles on political institutions and practices; and their contribution to changes in political structure (Tarrow 1998: 164). The central question raised by new waves of social protest is whether they are creating a transnational movement culture that threatens the structure and sovereignty of the national state (Tarrow 1998: 164).

Ethnoreligious cyberconflicts and the argument for moving towards network forms of organization

This idea has been fundamental in Arquilla and Ronfeldt's works on conflict in the information age. They basically argue that hierarchies have a difficult time fighting networks (e.g. Colombia, Algeria, the Zapatistas). It takes networks to fight networks and whoever masters the network form first and with the most success will gain major advantages (Arquilla and Ronfeldt 2001: 55). What Arquilla and Ronfeldt argue is that terrorists will continue moving from hierarchical to information-age network designs and that within groups 'great man leaderships will give way to flatter decentralized systems. This way more effort will go into building arrays of transnationally internetted groups than into building state-alone groups' (in Lesser *et al.* 1999: 41). As a result, power seems to be migrating to non-state actors, who are able to organize into 'sprawling multiorganizational networks', which are more flexible and responsive than hierarchies in reacting to outside developments and are better than hierarchies at using information to improve decision-making (in Lesser *et al.* 1999: 45).

This vision emphasizes adapting to a major consequence of the information revolution – the rise of network forms of organization. Especially after 9/11 it has become essential for states to become more flexible, in order to be able to face network-style organizations. Arquilla and Ronfeldt's thesis is that the challenge is

> to develop *hybrids* in which 'all channel' networks are fitted to flattened hierarchies. The major benefits may accrue in the areas of interagency and interservice cooperation. Since militaries must retain hierarchical command structures at their core, their hybrids should retain – yet flatten – the residual hierarchy, while allowing dispersed maneuver 'nodes' to have direct, all channel contact with each other, and with the higher command.
>
> (1997: 440)

The information revolution is favouring and strengthening networked organizational designs, often at the expense of hierarchies. States need to wake up to this fact and realize that networks can be fought effectively only by flexible network-style responses. This argument will be put to rest temporarily, to be addressed again in the discussion of information warfare in the section 'Information warfare' (pages 95–108) in Chapter 3.

3   The environment of cyberconflict

By the turn of the century a new kind of conflict named 'cyberconflict', describing conflict in computer-mediated environments (cyberspace), had emerged and become prevalent. This thesis seeks to introduce the key terms and themes of cyberconflict and argue that two different types of conflict occur: one between ethnic or religious groups fighting in cyberspace, as they do in real life; and another between social movements and their antagonistic institutions (hacktivism). This chapter unfolds the environment of cyberconflict by analysing the terms involved and engaging with the current debates in information warfare and security. More specifically, it includes analyses on hackers, information warfare, cyberterrorism, internet security analysis and cyberconflict's sociopolitical implications.

The political use of the internet has created a new lexicon, spawning terms like 'cyberwar', 'cyberattack' and 'netwar'. The term 'cyberconflict' (CC) – the generic reference to a certain form of politics on the internet – is now in regular use, but it has not yet been sufficiently clarified. This is because there are problems in defining and categorizing the wide variety of events occurring in cyberspace that fall under this conceptual umbrella. Here, the term cyberconflict is used to refer to conflicts of the real world spilling over to cyberspace. Typical of cyberattacks is the use by opposing parties of either Information Technology as such or IT as a weapon – for example, worms, Distributed Denial of Service attacks (DDoS), Domain Name Service attacks (DNS) or unauthorized intrusions – to attack the other side. The argument of this thesis is that cyberconflict includes two different categories of cyberpolitical action. These categories are sometimes blurred, but need to be distinguished in order to understand internet politics.

Netwarriors: terrorists or social activists?

It is not surprising that the internet has been used vigorously by social activists and campaigners all over the world. The internet quickly puts information into the hands of organizers, allows rapid replication of a successful effort, allows users to select their level of activity and helps publicize the campaign. It is therefore an organizing tool *par excellence*, because the more traditional tele-

phone trees or fax machines are too slow and the physical distances are too difficult and too expensive to cover (Danitz and Strobel 2001: 162). However, the
internet is not only used by social activists. An examination of historical precedents indicates that major political and military conflicts are increasingly
accompanied by a significant amount of online aggressive activity. Ongoing
conflicts also show that cyberattacks are escalating in volume, sophistication and
coordination (Vatis 2001). Parties in cyberconflicts have been described as terrorists or social activists depending on the discursive mood of their critics. This
is why it is important to examine the politics of this phenomenon and understand
its implications for future conflicts.

Information warfare

Hackers

It is vital to include an analysis of the hacking phenomenon and a discussion of
where it fits in the political environment of cyberconflict. Essentially, those
involved in ethnoreligious or sociopolitical environments of cyberconflict are
hackers – excepting of course social movement actors using the internet as a
mobilizational resource. I will briefly focus on Paul Taylor's (1999) influential
work *Hackers*, which provides excellent interviews and discussion on hackers,
computer scientists and computer security practitioners.

In contrast to my analysis of computer-mediated conflict between opposing
sides of a 'real world' conflict, Taylor concentrates upon the conflict between
the computer security industry and the computer underground (hackers). The
computer cognoscenti are split into two camps: the 'doves' who are prepared to
cooperate with the computer underground, arguing that hackers represent an
important stock of technical knowledge that society should not prematurely
isolate itself from by adopting a 'punish first, ask questions later' approach; and
the 'hawks', who advocate that the computer underground should be punished in
the courts (Taylor 1999: xi). Similarly, a definition of a 'hacker' varies, according to the discursive mood of the speaker:

> . . . to the lay person the phrase is likely to conjure up sensationalized images
> of malicious computer geeks in darkened rooms obsessively typing away;
> meanwhile to the computer aficionado, the phrase is more likely to be associ-
> ated with its dramatic fictionalization in the movies and the postmodern liter-
> ary genre of cyberpunk; to the computer programmer the term may refer to
> some of the earliest and most imaginative people involved in programming;
> and finally, wihin the computer security industry, the term *hacker* is likely to
> present a cue for opprobrium to be directed at 'electronic vandals'.
>
> (Taylor 1999: xii)

The word 'hacker', coined at MIT in the 1960s, connoted a computer virtuoso, who, according to the *New Hacker's Dictionary*, 'enjoys exploring the

details of programmable systems and how to stretch their capabilities; one who programs enthusiastically, even obsessively'. The meaning has evolved from the highly skilled but playful activity of academic computer programmers searching for the most elegant programming solution to being increasingly associated with its present day connotation of illicit computer intrusion (Taylor 1999: 13–14). Today's generation-x anarchist hackers share with their artisan and activist hacker predecessors a distrust of authority, a libertarian attitude and a tendency to position themselves outside bourgeois society's norms and values (Taylor 1999: 24, citing Hannemyr 1997). According to Taylor, the following are the main characteristics of a hacker: a) simplicity (the act has to be simple, but impressive); b) mastery (the act involves sophisticated technical knowledge); c) illicitness (the act is against the rules) (1999: 15). Thus, a hacker is defined not just by what he does but how he does it. The following are motivations for hacking: feelings of addiction, the urge of curiosity, boredom with the educational system, enjoyment of a feeling of power, peer recognition and political motivations. The computer industry is accused of over-emphasizing the vandal-oriented motivations and pathological aspects of hacking (Taylor 1999: 46). So how does hacking feel?

> ... the adrenaline rush I get when I'm trying to evade authority, the thrill I get from having written a program that does something that was supposed to be impossible to do, and the ability to have social relations with other hackers is all very addictive ... For a long time, I was extremely shy around others, and I am able to let my thoughts run free when I am alone with my computer and a modem hooked up to it ... If I were ever in a position where I knew my computer activity was over with the rest of my life I would suffer withdrawal.
>
> (email interview with Maelstrom in Taylor 1999: 48)

> Part of it was a sense of power. You were running an informal network of about 250 computers and no-one else outside your close circle of friends knew about it. The final goal was total world domination, to have everything under control. It was the ultimate game on the ultimate scale. You got a thrill out of knowing how much power you had. It was possibly hitting back at society. There was a sort of political anarchism involved. The main thrill was beating the system.
>
> (Taylor 1999: 56, citing Bowcott and Hamilton 1990)

In *Cybercrime: Law Enforcement, Security and Surveillance in the Information Age*, Thomas writes that what is talked about in terms of hackers is the manner in which hackers themselves exist in a shadow space of secrecy, possessing near-mystical powers that allow control of technology that itself is beyond discourse. The hackers themselves are coded in such a way that they literally become the secret that needs to be broken. Fundamentally, one can watch a hacker's actions, even monitor them online, but this means nothing until they can be attached to a real body and therefore become prosecutable. A discussion about hackers is important to this work because individuals and groups engaged in sociopolitical and ethnoreligious cyberconflicts are part of the hacking culture in a particularly politicized way.

Nowadays, hackers fight for clearer political goals than their forebears, especially when campaigning for information freedom. Many hackers believe that electronic communications are unsafe, with governments legally tapping data lines, copying electronic mail and suspecting hacking often enough to get search warrants or to confiscate equipment. Accordingly, Goggans, another hacker Taylor interviewed, argues:

> I know all too well how simple it is to view and alter consumer credit, to transfer funds, to monitor telephone conversations etc ... I can monitor data on any network in existence, I can root privileges on ANY Sun Microsystems UNIX. If I, a 22 year-old, non-degreed, self-taught, can do these things, what can a professionally taught profit motivated individual do? THERE IS NO PRIVACY ... People need to know the truth about the vulnerabilities of the computers they have entrusted their lives to.
>
> (in Taylor 1999: 70)

However, as Taylor and Jordan argue, 'hackers' over-identification with technical means over political ends and their parasitic relationship to various technological systems means that although they are at the heart of the exercise of power, they remain in an ultimately powerless dependent relationship' (2004: 162).

Further, the computer underground is anarchic, a confederation of phreakers/hackers and virus writers from all over the world whose common interests transcend culture or language. 'They change IDs, aliases, sites, their methods and gang membership as rapidly as the authorities track them. Stamping out hacking is like trying to pin down mercury' (Taylor and Jordan 2004: 28, citing Clough and Mango 1992: 18).

Hacking can potentially perform a variety of benevolent services to the security industry: it has been responsible for many of the most progressive steps in software development; it reflects the ways in which the development of high technology has outpaced orthodox forms of institutional education; it is an important form of watchdog counter-response to the use of surveillance technology and data-gathering by the state, and to the increasingly monolithic communications power of giant corporations; and finally, as guerilla know-how, it is essential to the task of maintaining fronts of cultural resistance and stocks of oppositional knowledge, as a hedge against a techno-fascist future (Taylor and Jordan 2004: 43, citing Ross 1991: 82). Also, the hacker presence constantly pushes forward the limits of computer security and thus contributes to a general climate of improved security consciousness, without which 'the nation would be nakedly awaiting serious attack from thieves and foreign agents' (Maelstrom interview in Taylor and Jordan 2004: 97).

On the other side of the coin, the business sector is reluctant to report computer crime for two reasons. First, for fear of adverse publicity, public embarrassment or loss of good will. Second, for fear of the loss of investor or public confidence and the resulting economic consequences. Official statistics on hacking incidents tend to be unreliable: first, because the organizations affected may be totally oblivious to security breaches that have occurred, leading to possible under-reporting; and second, because there is a tendency for both computer security figures and hackers to hyperbolize the situation, leading perhaps to over-reporting. At the same time, serious security incidents are whitewashed, to prevent the management from being accused of negligence (Taylor and Jordan 2004: 67).

One of Taylor's main arguments is that a knowledge gap exists between the computer security industry and the computer underground. Due to the fact that computer security is lacking in any theoretical grounding, the only true way to test security is to actually attempt to breach a system. There is a scarcity of theoretical knowledge surrounding computer security with various calls for more hands-on experience of security to supplement more formal theory. 'The knowledge gap is thus rooted in the difference between theoretical concepts and guidelines to security and the "nitty gritty" of real world computing situations where security weaknesses flourish in the interstices of continually expanding and evolving computer systems' (Taylor and Jordan 2004: 79). Hackers are sometimes used to conduct the testing. The problem, however, is that hackers may have potentially useful knowledge, but such knowledge often does not sit comfortably with the academic and commercial worlds' preference for ethically unproblematic and rigorously researched knowledge. Often, hackers are labelled as deviant, despite the fact that the hacker community shares some of the same characteristics as its computer security industry counterpart.

These are tough times for hackers. Federal agencies now have broad new powers to spy on them, thanks to provisions in the anti-terrorist USA Patriot Act. The House of Representatives in the United States passed a law that could send hackers to prison for life. On top of that, former hackers have difficulty finding jobs, because they must compete against a surplus of people with similar skill levels.

Information warfare and cyberterrorism

The most common view on information warfare and the future of conflict, whose best-known exponents are Heidi and Alvin Toffler (1980), extrapolates from the idea that territory, population and natural resources are becoming less important, relative to human capital and the possession of information. Taking this process to its logical conclusion, these theorists believe that information will soon become the key source of wealth and power – equivalent to steel, coal and oil in the industrial age, or fertile land in the agricultural age. This change will eventually amount to a social revolution, whose scope is equivalent to only two previous such transformations: the agricultural and industrial revolutions

(Shapiro 1999: 117). Just as the transition from agriculture to industry was correlated with the industrialization of warfare, so too counterparts aver that people war as they work (Owens 1995). The transition will be from industry to information-based services and this will correlate with the 'informatting' of warfare. War waged in cyberspace might be bloodless and even clean, a possibility that has led one high-ranking military officer to see information technology as 'America's gift to warfare' (Owens 1995). Sun Ji is an icon in this pantheon, with his observation that the 'acme of skill' consists in winning without fighting. Advanced technological systems will not only help shape the environment of future conflict, but will also magnify the importance of the psychological battle to the conflict outcome.

One definition of cyberterrorism is 'computer-based attacks intended to intimidate or coerce governments or societies in pursuit of goals that are political, religious or ideological' (Denning 2001). So, who can be a cyberterrorist? The answer is anyone, David and Sakurai argue, citing a CERT/CC April 2002 report, which states the level of automation in attack tools continues to increase. Automated attacks commonly involve four phases: (1) scanning for potential victims; (2) compromising vulnerable systems; (3) propagating the attack; (4) coordinated management of attack tools (David and Sakurai 2003: 15–26).

Indeed, the problem for American defence experts is that they don't know who the enemy is:

> Whether they are disgruntled Americans, Hamas terrorists or pariah dictators as Saddam, the attackers could wage cyber warfare undetected on any laptop computer from the Sinai desert to Singapore. Just as exasperating for the government would be deciding how to deploy its military. 'If you don't know who your enemy is, how can you retaliate?' said one expert. This makes cyber warfare the great equaliser, a cheap and effective weapon for any third world rogue state or small terrorist organization wanting to wage war against a super-power and win. All they might need is a few million dollars to hire a handful of 'cyber mercenaries' capable of penetrating supposedly secure systems.
>
> (*Sunday Times* 17 May 1998: 26)

Michael Vatis, director of the Institute for Security Technology Studies at Dartmouth, reported the following on 26 September 2001 before a US House Subcommittee:

1 the likelihood of cyberattacks against the US and allied information infrastructure is high;
2 such attacks could come from terrorists and/or their nation-state sponsors, but are more likely to come from sympathizers of terrorists or of nation states targeted by US military operations;
3 such attacks will almost certainly target the websites of government agencies and private companies in the US and allied countries, but could

also attack more high-value targets such as networks that control critical infrastructures;

4 such attacks could utilize destructive worms and viruses, Distributed Denial of Service exploits, and intrusions to disrupt targeted networks;

5 such cyber exploits could be combined into a potent mix to cause widespread disruption, and also combined with physical terrorist attacks to maximize the destructive potential of both sets of terrorist tools.

(in David and Sakurai 2003: 15–26)

The concept most commonly used to connect IT and the military is information warfare. What this means is that the object of conflict is no longer territory or resources, but information. The object of information warfare is the control over information as a source of wealth and power (Shapiro 1999: 119). The US Air Force has described it as encompassing 'any action to deny, exploit, corrupt or destroy the enemy's information and its functions, protecting ourselves against those actions and exploiting our own military information functions' (US Air Force 19 December 1994). The battlespace of information warfare is cyberspace – an ethereal place which does not fit neatly into the land–sea–air space metaphor, where conflicting parties strive for information superiority and dominance. According to Shapiro, 'information dominance means that your side has the ability to collect, communicate, and protect information without disruption, while the other side does not' (1999: 130). Taking out all information-transfer media would bring down a country's stock market, banking system, air traffic control, emergency dispatches and more. In November 2003, the US Homeland Security Department's first simulation of a terrorist attack on computer, banking and utility systems exposed problems with the ways victimized industries communicated vital information during the crisis. It simulated physical and computer attacks on banks, power companies and the oil and gas industry. Electronic communications may have been used by the September 11 terrorists, and are certainly used by some of those accused of related terrorism offences, such as alleged 9/11 accomplice Moussaoui, who requested that prosecutors turn over their records of messages sent and received through his email account, claiming his email could help him establish his alibi (Delio 30 August 2002).

Remarkably, terrorists are being bred and trained to be technologically savvy and to attack, disrupt, damage and perhaps even destroy technology infrastructures and computer-based economic activities. They are the biggest threat to the information infrastructure and the new digital economy (Erbschloe 2001: 37). Cyberterrorists could act in the name of one religion or another and support or be supported by one or more outlaw nations. They will attack anything that is vulnerable and will focus their attacks on headline-grabbing efforts to make civilian populations fearful and to embarrass government officials and organizations. Erbschloe has offered the following conclusions regarding terrorists' use of information warfare:

Because of the international nature of information warfare, existing political structures are not prepared to immediately investigate the source of attacks and to appropriately respond to and counter information warfare scenarios. It is not likely that all governments will equally participate or cooperate in the investigation or information warfare attacks or that they will work together to halt attacks ... Information warfare attacks can effectively impact the economics of a nation or region without destroying infrastructures or disabling military capabilities.

(2001: 95)

According to Erbschloe, sustained terrorist information warfare strategies are the ongoing deliberate efforts of an organized political group against the military, industrial, civilian and governmental economic information infrastructures or activities of a nation, region, organization of states, population or corporate entity.

In purely information-warfare terms, there are five dynamics that favour the outlaw attacker against the defender (Erbschloe 2001: 174–175). First, a fundamental dynamic of information warfare is that defenders must always succeed in protecting systems, whereas if attackers do not succeed, they can try again or move to another target. A second dynamic is the growth of computer networks and the increase of internet connectivity. A conversation with a director of computer security at a large telecommunications firm reveals the magnitude of the problem. When asked, 'Are all of the computers in your company secure?' the answer was: 'They will be when I find out where they all are' (Erbschloe 2001: 174). A third dynamic is that the attackers have access to all of the same technology the defender has, as well as all of the technical information about systems, including weaknesses in hardware and software. On a related note, attackers can use the internet and become members of the same clubs, chat rooms, bulletin boards and email lists that defenders use to get information about products and to talk with their peers. Lastly, attackers can not only work from almost anywhere in the world, but they can easily outnumber defenders. Also, hackers and computer bandits take as much pride in their abilities and accomplishments as do the people trying to stop them. Similar levels of pride have been noted for terrorists, especially those who have a religious basis for their actions. Ultimately, as Erbschloe argues, religious or politically motivated terrorists usually gain little in the way of financial compensation, but are often revered in their community and can die as heroes in their countries (2001: 259).

In such an arena, several questions emerge. First, is a strategic information attack tantamount to a physical attack? Second, can a military advantage in information technology guarantee success within low-intensity conflict, against low-tech asymmetrical strategies (Gombert 2001: 61)? Third, is there a transition from platform-centric warfare (wherein networks exist to enhance platform performance) to network-centric warfare (wherein platforms are the eyes, ears and fists of a broader entity) (Libicki and Shapiro 2001)? Although it is too early to answer, there are some early indications. As far as Shapiro is concerned, there will be a definite change in battle dynamics: 'commanders would no longer be encouraged to use intuition or take risks because "dominant battle space knowledge" would render such attributes unnecessary or even dangerous. A radically different military and a very different style of fighting would certainly result' (1999: 146). Also, command, control, communications, computing, intelligence, surveillance and reconnaissance (C41SR) will control the outcome of the conflict (Gombert 2001: 59).

At the systemic level, information warfare is the organization of information to provide warriors with what has been termed 'dominant battlespace knowledge', an important component of which is the American Department of Defense's nascent 'system of systems'. Insofar as the ability to kill what can be seen makes seeing (locating, identifying and tracking) the key to war, seeing is increasingly best done by networking sensors and human observers to create a shared foundational truth that forms the basis of command, control and operations.

One of the most influential centres analysing information warfare is the Rand Corporation. Starting with their work *Countering the New Terrorism*, Lesser, Hoffman, Arquilla, Ronfeldt, Zanini and Jenkins introduced the concept of cyberterrorism and, in subsequent publications analysed below, gave a thorough analysis of the issues involved.

Terrorism, according to these authors, has always been about information:

> ... from the fact that trainees for suicide bombings are kept from listening to international media, through the ways that terrorists seek to create disasters that will consume the front pages, to the related debates about countermeasures that would limit freedom of the press, increase public surveillance and intelligence gathering, and heighten security over information and communication systems.
>
> (Lesser *et al.* 1999: 72)

According to a report released on May 1999 by the Rand Corporation, which was the result of a year-long project sponsored by the Airforce's Deputy Chief of Staff for Air and Space Operations, there will be a new form of terrorism known as 'netwar':

> The rise of networks is likely to reshape terrorism in the Information Age and lead to the adoption of netwar – a kind of Information Age conflict that will be waged principally by nonstate actors. There is a new generation of radicals and activists who are just beginning to create Information Age ideologies. New kinds of actors, such as anarchistic leagues of computer-hacking 'cyboteurs' may also partake of netwar.
>
> (Verton 3 May 1999)

The same report advised that the US Air Force should slow its modernization plans and rethink its connections to the internet if it wants to fight off a revolutionary if as yet undeveloped form of cyberterrorism. The rise of networks is likely to reshape terrorism in the Information Age and lead to the adoption of netwar – a kind of Information Age conflict that will be waged principally by non-state actors (Verton 3 May 1999).

The Rand report also predicts that cyberterrorists will use new tactics such as 'swarming', which occurs when members of a terrorist group, spread over great distances, electronically converge on a target from multiple directions, a tactic different from the traditional form of attacking in waves, which delivers a knockout blow from a single direction on the internet (Arquilla and Rofeldt 2000).

However, it is useful to stress that terrorists might be more interested in keeping the internet up and running than in disrupting or destroying its components. One has to agree with Arquilla and Ronfeldt that network-based conflict will become a major phenomenon of the future, as can be clearly demonstrated with al-Qaeda long before the September 11 attack, with reports of bin Laden having advanced information systems designed by Egyptian computer scientists (2001). As Arquilla and Ronfeldt argue:

> These protagonists are likely to consist of dispersed small groups who communicate, co-ordinate and conduct their campaigns in an internetted manner, without a precise central command ... To give a string of examples, netwar is about the Middle East's Hamas more than the Palestine Liberation Organization (PLO), Mexico's Zapatistas more than Cuba's Fidelistas, and the American Christian patriot movement more than the Ku Klux Klan.
>
> (2001: 45)

Arquilla and Ronfeldt argue that power seems to be migrating to non-state actors, who are able to organize into 'sprawling multi-organizational networks', which are more flexible and responsive than hierarchies in reacting to outside developments, and appear to be better than hierarchies at using information to improve decision-making (Lesser *et al.* 1999). Essentially, what these writers argue is that conflicts will evolve around 'knowledge' and the use of 'soft power'. This will come about with the help of information-age ideologies, in which identities and loyalties shift from the nation-state to the transnational level of global civil society.

When Lesser *et al.* talk about networks, they describe three types: a) the chain network where actors move along a chain of separated contacts, and where end-to-end communication travels through the intermediate nodes; b) the star, hub or wheel network, where a set of a actors is tied to a central node or actor, and must go through that node to coordinate; and c) the all-channel network, as in a collaborative network of small, militant groups where every group is connected to every other (Lesser *et al.* 1999: 49). It is important to note that netwar is not simply a function of the internet; it does not take place only in cyberspace or the infosphere. As Lesser *et al.* argue:

> Some key *battles* may occur there, but a *war's* overall conduct and outcome will normally depend mostly on what happens in the real world. Even in

information-age conflicts, what happens in the real world is generally more important than what happens in the virtual worlds of cyberspace or the infosphere. Netwar is not Internet war.

(1999: 53)

For example, Hamas uses internet chat rooms and emails in the United States to coordinate their activities across Gaza, the West Bank and Lebanon, making it difficult for Israeli security officials to trace their messages and decode their contents (Denning 2001). In a 1998 US News and World Report there was evidence of 12 of the 30 groups on the US State Department's list of terrorist organizations on the web. More recently, it seems that virtually every terrorist organization is on the internet (Denning 2001: 252). For instance, the presence of Middle Eastern terrorist organizations on the internet is suspected in the case of the Islamic Gateway, a world wide website that contains information on a number of Islamic activist organizations based in the United Kingdom. British Islamic activists use the world wide web to broadcast their news and attract funding; they are also turning to the internet as an organizational and communication tool. While the vast majority of Islamic activist groups represented in the Islamic Gateway are legitimate, one group – the Global Jihad Fund – makes no secret of its militant goals (Lesser *et al.* 1999: 66).

Moreover, an internet site claiming to represent al-Qaeda said that the network decided to launch suicide attacks against Israel with the goal of destroying the Jewish state. US government officials believe that the site indeed speaks for al-Qaeda, and intelligence officers had been monitoring it for some time. Bruce Hoffman, a terrorism expert with the Rand Corporation research group, commented that al-Qaeda's new attacks on Israel stem from terrorists looking for work, with al-Qaeda desperate to appear relevant, to be a player in Middle Eastern politics (Mintz). Also, Rep. Lamar Smith (R-Texas) warned in closed-door briefings for members of Congress that there is a 50 per cent chance that the next time al-Qaeda terrorists strike the United States, their attack will include a cyberattack. Smith said that officials from federal law enforcement and intelligence-gathering agencies disclosed that al-Qaeda operatives have been exploring US websites and probing the electronic infrastructure of American companies in search of ways to disable power and water supplies, disrupt phone services and damage other parts of the critical infrastructure (Matthews 25 July 2002). In more subtle terms, Azzam Publications, based in London and named after Sheikh Abdullah Azzam, a mentor of Osama bin Laden, has a site dedicated to Jihad around the world. It is alleged that the site, which sold Jihad-related material from books to videos, was raising funds for the Taliban in Afghanistan and for allied guerrillas fighting the Russians in Chechnya (Miami Herald 3 April 2001).

Also, US forces hunting down al-Qaeda and Taliban fighters strung fibre-optic cables through the treetops and installed satellite uplinks, providing thousands of American troops with a crucial link to home: the internet. The technology, of course, stands in stark contrast to the spartan surroundings, as few people in Afghanistan have phones and fewer still have ever seen the internet (USA Today 5 July 2002).

Jerry Everard, in his work *Virtual States*, argues that a developed nation engaged in open conflict with a small, less-wired nation could find its own economic system targeted, with its banking system, its stock exchanges, its telecommunications and power grid systems, and its logistical support networks being systematically targeted from any point on earth. 'Such activities could result in swaying the hearts and minds of the domestic polity of the developed west toward the conclusion that big states may find the cost of small wars far greater than they bargained for' (Everard 2000: 115).

Advanced technological information systems will allow state and substate actors, including news services, non-governmental organizations and even individual citizens, to make voice, video and written information instantly available to audiences located in the remotest areas of the globe. This touches on another important factor: information classification. Up until now, information data in the US was put into separate partitions, where only a few people had access to it.

What is needed for effective governance, however, is for authorities to make information available 'by job' to whoever needs it – regardless of their security clearance. 'We have to change the way we classify information. The old system may have worked against the Soviet Union. But today, the federal government needs to make information available to law enforcement, and the security staff guarding the power plant', Jim Caverly, who heads the Homeland Security Department's Information Analysis and Infrastructure Protection division, has commented (Shachtman 16 October 2003).

Steven Aftergood, who heads the Federation of American Scientists' Project on Government Secrecy, has also pointed out that 'any attempt to control the flow of information impedes the whole. It's the difference between a top down command structure and a network' (Shachtman 16 October 2003). Military and intelligence officials are trying to move away from their old hierarchies and towards a structure in which every soldier, every drone and every general is connected by computer networks. As Jordan Crandall writes:

> Computerization has brought massive changes in the development and coordination of databases, the speed and quality of communication with intelligence and tactical agencies, operations and combat teams. New technologies of tracking, identification, and networking have increased this infrastructure into a massive machinery of proactive supervision and tactical knowledge. Originally conceived for the defense and intelligence industries, these technologies have, after the cold war, rapidly spread into the law enforcement and private sectors.
>
> (15 June 1999)

Furthermore, due to the fact that netwar defies and cuts across standard boundaries and jurisdictions, governments have difficulty assigning responsibility to a single agency (military, police or intelligence) to respond. This poses a challenge to the state, because traditionally, ideals of sovereignty and authority are 'linked to a bureaucratic rationality in which issues and problems can be neatly divided, and specific offices can be charged with taking care of specific problems. In netwar, things are rarely so clear' (Crandall).

Mulveron contemplates various scenarios in *The People's Liberation Army in the Information Age*, assessing the capabilities of the US and China in a possible confrontation. He writes that there are important differences between Chinese and American Information Warfare (IW) literatures. People's Liberation Army (PLA) writers look at IW in strictly military terms, while Western authors accept the dichotomy between information warfare waged between states or militaries (cyberwar) and information warfare waged between substate actors and states (netwar). Both US and Chinese authors are guilty of over-using Sun Ji, especially the notion of 'winning the battle without fighting'. Chinese theorists are also forced to discuss from a technologically inferior standpoint, in opposition to an advanced foe (Mulveron and Yang 1999: 182). Nevertheless Mulveron believes that IW presents the Chinese with a potentially potent, if circumscribed, asymmetrical weapon. 'Defined carefully, it could give the PLA a longer-range power projection capability against US forces that its conventional forces cannot currently hope to match ... to attack its information systems, especially those related to command and control and transportation' (Mulveron and Yang 1999: 175).

The reason for this is that launching a cyberattack is fairly inexpensive. One knowledgeable hacker with a computer can wreak havoc on an automated pipeline. A cell of 'cyber-space guerrillas' armed with a few thousand dollars' worth of hardware could disable a nation's power grid. Several hackers together can dramatically increase the capabilities of a terrorist group (Jane's International Police Issues 2 December 1998). More guerrilla groups will be attracted to cyberwarfare, because they can spread propaganda, recruit sympathizers and collect data. The possibilities are endless: military espionage, control and disruption of information flow, destruction, distortion and fabrication of data, electronic bombs and psychological operations could be potential tactics.

On the other hand, Winn Schwartau, author of *Pearl Harbor Dot Com*, argues against those who believe that information warfare is just hype, that cyberattacks can cause terror and that the fundamental flaw in the sceptics' argument is their assumption that, because things haven't gone wrong before, they will never go wrong. It relies on a flawed fortress mentality, which has never worked as a defence (Shachtman 20 December 2002). Shachtman gives an example of viable electronic terrorism, which would involve an IW attack against a series of US domestic air carriers, where the attackers do not use a conventional bomb but, rather, an electromagnetic bomb, which sufficiently interferes with the avionics of the plane to cause it to crash (Venke 4 August 1996).

As early as 1995, the Pentagon's assistant director for strategic planning made clear that the problem is real enough:

> ... as the information age matures, a truly revolutionary form of warfare will emerge. Information warfare will be fought in a different environment,

with adversaries grappling in cyberspace. As every potential adversary achieves access to multiple information systems, warfare will be conducted virtually at the speed of light over global distances. Domination of cyber-space may render the need to employ conventional forces and firepower less likely.

(Guisnel 1997)

Consequently, there must be some perceived threat, as the US government has devoted dozens of committees to deal with information warfare. The most important ones are the National Information Protection Center (NIPC), consisting of FBI agents and military and national security components, the Critical Infrastructure Coordination Group and the US Commerce Department's Critical Infrastructure Assurance Office (CIAO). The latter include business and private-sector specialists (Pence Wired). The special force is the Air Force Information Warfare Center (AFIWC) at Kelly Air Force Base in San Antonio, which monitors activity in the nation's defence networks and in case of attack goes on the offensive. However, in January 2003, the US government issued a revised cybersecurity plan, reducing by nearly half its initiatives to tighten security for vital computer networks, placing more emphasis on the New Homeland Security Department and eliminating the plan to consult regularly with privacy experts. The plan was to launch some tests against civilian US agencies and improve the safety of automated systems that operate the nation's water, chemical and electrical networks, while also reserving the right to wage cyberwarfare if the nation is attacked (Associated Press 7 July 2002). 'When a nation, a terrorist group, or other adversary attacks the United Sates through cyberspace, the US response need not be limited to criminal prosecution', the plan said. Also, the US government asked internet users and businesses to practise 'safe computing online' and promised to bolster its own cyber-defences (Associated Press 7 July 2002).

However, this National Strategy to Secure Cyberspace imposes few new requirements on the private businesses that control 85 per cent of the global computer network. 'Instead the Bush administration sees its role largely as a cheerleader, encouraging businesses to keep their networks secure and supporting publicity campaigns to encourage greater individual use of anti-virus software, firewalls and other security tools' (Reuters 19 February 2003). The Department of Homeland Security is expected to lead a response when cyberattacks occur, set up programs to develop a more tech-savvy workforce, and encourage business sectors like banking and utilities to bolster security standards on their own.

The full extent of the US cyber-arsenal is among the most tightly held national security secrets, even more guarded than nuclear capabilities. Because of secrecy concerns, many of the programs remain known only to strictly compartmented groups, a situation that, in the past, has inhibited the drafting of general policy and specific rules of engagement. As Major General James David Bryan, who heads the Joint Task Force on Computer Network Operations, explains, his group has three main missions: to experiment with cyber-weapons in order to better understand their effects; to normalize the use of such weapons, treating them 'not as a separate entity', but as an integral part of the US arsenal; and to train a professional cadre of military cyberwarriors (Graham 7 February 2003).

Furthermore, McCaffrey, a highly decorated combat veteran, told attendees at the American Society for Industrial Security (ASIS) annual conference that the government's ability to protect the country is 'only as good as the technology that backs it up' (Graham 7 February 2003). The government's current snooping system, known as Carnivore, makes it too easy to enable the reading of all emails with only a warrant. As a result, indiscriminate access makes it difficult for local law-enforcement agencies to find useful evidence in a sea of data. Kelly Kuchta, a cybersecurity expert who is chair of ASIS's information technology security council, sees private security firms as being more willing to work with law enforcement agencies since September 11, 2001, sharing information about cyberattacks with the FBI as part of InfraGuard, a cooperative program between the public and private sectors (Gartner 11 September 2002). The Bush administration has toughened anti-hacking laws since September 11 and increasingly lobbied governments to cooperate in international computer-crime investigations. The United States and Britain were among 26 nations that last year signed the Council of Europe Convention on Cybercrime, an international treaty that provides for hacker extraditions even among countries without other formal extradition agreements.

Such zeal on behalf of national governments could be explained, if we consider the economic, military and socio-political aspects of cyberconflict. The protection of commercial information infrastructure is a national security concern, which cannot be disregarded easily. According to a study conducted by the FBI, nearly 90 per cent of US businesses and government agencies suffered hacker attacks in 2002, only a third of companies that suffered attacks reported the intrusions to law enforcement and around 78 per cent of companies surveyed also said that employees had abused their internet access privileges by downloading pornography or private software (Newsbyte 17 June 2002).

The burden is therefore on the government to demonstrate that the protection of commercial information infrastructure is a national security concern that cannot be discharged any other way. Convincing a population wary of government intervention of the need for such intrusive government action may require a crisis such as September 11.

Internet security analysis: incidents and responses

Incidents and responses

An examination of historical precedents indicates that major political and military conflicts are increasingly accompanied by significant cyberattack activity. Ongoing conflicts also show that the cyberattacks are escalating in volume, sophistication and coordination (Vatis 2001).

It would be useful here to include some cases of internet security breaches. As early as March and April 1994, Air Force computer security experts discovered that their classified network at the Rome Laboratories in New York had been attacked. Attackers gained complete access to all Rome networks, copying sensitive (though not classified) information, and while they could have brought the network down, they chose not to (Shapiro 2001: 132). Through the Rome Lab they accessed other classified sites like the South Korean Atomic Research Institute. The hackers were discovered as codenamed 'Kuji' and 'Datastream Cowboy'. Datastream was identified through an informant after he bragged and left his phone number with him. He turned out to be a 16-year-old using nothing more than a 486sx PC (Selden November 1996). In 1994, they were arrested in England, and turned out to have planted eavesdropping software that allowed them to monitor emails and other sensitive information (Associated Press 12 November 2002).

In one long-running operation, the subject of a US spy investigation dubbed 'Storm Cloud' and 'Moonlight Maze', hackers traced back to Russia were found to have been quietly downloading millions of pages of sensitive data, including one colonel's email inbox. Most recently, in April 2001, government computer operators watched as reams of electronic documents flowed from Defense Department computers, among others. Remarkably, Gary McKinnon, a British hacker indicted for hacking into scores of US military computers, installed copies of a commercial remote access utility called RemotelyAnywhere on navy and other military systems he hacked. For nearly a year, McKinnon was able to control a vast network of defence computers without detection, authorities said. Using a personal computer connected to an ISP in England, McKinnon downloaded a trial copy of RemotelyAnywhere in March 2001 from a server maintained by Binary Research, the Milwaukee-based distributor of RemotelyAnywhere. To obtain a special code to unlock the demonstration software, McKinnon also provided his girlfriend's email address. McKinnon was indicted in federal courts in Virginia and New Jersey on eight counts of computer crimes. McKinnon was facing on each count a maximum sentence of ten years in prison and a $250,000 fine. Investigators characterized the hacker as 'a conspiracy theorist' who 'seemed to think that the government was controlling all sorts of things' (McWilliams 15 November 2002).

A more publicized incident was the Slammer worm, also known as Sapphire, which spread in ten minutes in the last weekend of January 2003, quickly crippling many computer systems around the world, offline. It targeted Microsoft's SQL Sever 2000, as well as applications created with the Microsoft SQL Server 2000 Desktop Engine. Microsoft had released a patch for Slammer in July 2002, but security experts said a successful installation required users to manually edit system files, a compilation that resulted in some patches being installed incorrectly, if they were installed at all. The worm infected 247,000 computers worldwide, while the code was exceedingly condensed and did not include references to hacker aliases or locations. It used a transmission method that made it especially easy for its author to throw off investigators, by falsifying his digital trail.

Many top experts believe the programming for the internet worm was based on software code published on the web months before by a respected British computer researcher David Litchfield, and later modified by a virus author within the Chinese hacker community known as Lion. Litchfield said he originally published the blueprints for computer administrators to understand how hackers might use the program to attack their systems ('Few clues in worm whodunnit', Associated Press). Another issue cropping up with the Slammer worm was that the internet security firm Symantec withheld information about the worm for hours after spotting it, sharing the information only with select customers, leaving the rest of the global community to cope with the impact of Slammer. South Korean telephone systems administrator Lee Ji-Ho, whose country's entire communications system collapsed for roughly 24 hours because of Slammer, said: 'my country was hurt hard by this Slammer. We could have been prepared and maybe avoided had we known ahead' (Delio 'What Symantec Knew but Didn't Say). Code Red and Nimda spread around the world in a matter of hours, but Slammer took under three minutes to affect thousands of machines, and was able to compromise nearly all vulnerable systems in about half an hour. Another computer virus found in the wild in October 2003, called Mimail.C, was capable of turning infected personal computers into 'spamming' machines (Reuters 31 October 2003).

As far as viruses are concerned, about 1,000 viruses are created every month by virus writers who increasingly intend to target new operating systems. Jan Rhuska, the chief executive of Sophos, the world's fourth largest anti-virus provider, has commented that the number of viruses created would continue to climb in the coming years and said of the virus writers, believed to be computer-obsessed males between the ages of 14 to 34 years, that 'have chronic lack of girlfriends, are usually socially inadequate and are drawn compulsively to write self-replicating codes' (Reuters 18 March 2003). In January 2003, Welsh virus writer and web designer Simon Vallor, 22, was sentenced to two years in jail for spreading three mass-mailing computer viruses that infected more than 27,000 computers in 42 countries. Some of the viruses like Klez, a mass-mailing worm that originated in November 2001, simply refuse to go away (Reuters 18 March 2003). Members of the computing industry and law enforcement, when testifying before the technology subcommittee of the House Committee on Government Reform in the US, proposed solutions for these kinds of attack, such as better standards for producing software, computing ethics education directed at children, increased funding and training for computer forensics to catch hackers and virus writers, and protocols for information sharing that would aid in capturing perpetrators across borders (Zetter 11 November 2003).

The year 2003 has been deemed the worst in computer-virus history by security experts, despite the fact that worm and virus writers displayed no significant technological progress in the code of their new creations. That year, computer worms managed to shut down ATMs, slow airline and train travel by affecting reservation and signalling systems, clog emergency phone services, and crash networks controlling critical systems at hospitals and at least one nuclear plant.

'The scary thing is that the Blaster and Slammer worms were not intended to take down critical systems. The worms were just programmed to propagate. Imagine what could have happened if they had carried a malicious payload', said Mikko Hypponen, director of anti-virus research for F-Secur (Delio 23 December 2004).

The most troubling development in 2003 was that spammers and virus writers have evidently decided to partner, resulting in a demon love-child virus-worm that both infects computers and spews spam. Slammer, released in January 2003, has been described by many as the biggest attack against the internet ever. Spreading over network connections, Slammer searched for victims so aggressively that it caused an enormous amount of network traffic, dramatically slowing network response times across the internet.

Following Slammer, the next big virus was Blaster (aka MS Blast), detected in August, which wormed into computers through a hole in the Windows operating system. Millions of Windows 2000 and Windows XP users were alerted to the presence of Blaster in their computers when they saw an error message informing them their machine would be restarting in 60 seconds. Just a week after that, Sobig.F was released, quickly becoming the worst worm ever, sending over 300 million infected messages around the world, according to anti-virus experts. 'It is likely that there is a virus-writer group behind Sobig. They used the worm to infect a huge number of computers and then sold various spammer groups lists of proxy servers, which would be open for spreading spam. It was clearly a business operation', commented Hypponen, a security expert with F-Secure. Experts expect that this type of commercial activity will increase, and say that viruses will quickly become spammers' tool of choice in their never-ending efforts to force junk mail into people's inboxes (Delio 23 December 2004).

The same pattern was noticeable in January 2004 with the MyDoom virus, whose social engineering specifically targeted the corporate world. 'Malicious hackers have been selling the use of compromised commuters to spammers for some time now – it's good money for them. Compromised home-user machines are already in widespread use by spammers and MyDoom-infected machines will only add to that pressure on inboxes', said Jose Nazario, security analyst for Arbor Networks (Delio 28 January 2004). Additionally, German police have arrested an 18-year-old suspected of creating the Sasser computer worm, which surfaced in May 2004 wreaking havoc on personal computers. 'Hopefully this arrest will limit their activities. If we can start catching these guys it will certainly put more pressure on existing virus writers' said Hyponnen (Delio 23 December 2004). Also, due to the fact that pieces of code found in a recent version of the Netsky worm made references to Sasser, Graham Cluely, consultant at Sophos, said the police may have cracked the Netsky gang with the arrest. The economic toll of Sasser may never be known, but it has claimed some big scalps, including Germany's Deutsche Post, Britain's coastguard stations and the investment bank Goldman Sachs (Reuters 8 May 2004).

The economic threat in cyberspace manifested itself when a Mafia-led syndicate used banking and telecommunications insiders to break into an Italian bank's computer network, diverting the equivalent of $115 million in European Union aid to Mafia-controlled bank accounts overseas before Italian authorities detected the activity (www.cia.gov). Additionally, in June 2003, a huge internet piracy ring was shut down in an Italian financial police operation that resulted in the arrest of 181 people and the recovery of pirated software worth about $139.5 million. The piracy network used email and illegal websites to market itself and distributed goods via mail order (Sturgeon 16 June 2003). Moreover, one of the most impressive cyberattacks occurred in October 2002, when nine of the 13 computer servers that manage global internet traffic were crippled. The attack lasted for an hour, but it was described as the most sophisticated, large-scale assault against these crucial computers in the history of the internet (Wired.com 23 October 2002).

Despite the fact that there have been several serious incidents of cyberattack, doubts have been raised as to the real or exaggerated nature of cyberwar. Several experts think that information warfare will likely go much further than nuisance attacks by hacktivists. In a rather humorous mood, the Crypt Newsletter wrote that 'the wonderful thing about secret cyberwar is that it can be anything anyone wants it to be. In secret cyberwar, it is not really necessary that anyone be an actual reliable witness to it or that effects of it even be presented or seen' (undated).

The point made is that there are instances where cyberwar stories are reported in the media, but not experienced by anyone else. The criticism the Crypt Newsletter made late in 1998 in *Issues in Science and Technology* magazine was that one has to be sceptical of the warnings of information warfare, because those alarms often come from the very people who are to benefit from government spending to combat the threat. An example they give to reinforce this argument is that the primary author of a January 1997 Defense Science Board report on information warfare, which recommended an immediate $580 million investment (and $15 billion over five years) in private sector R&D for hardware and software to implement computer security, was Duane Andrews, executive vice-president of SAIC, a computer security vendor and supplier of information warfare consulting services (Sun.soci.niu.edu/~crypt).

Another instance of absurd claims was criticized by the Crypt when *The Economist* reported a Pentagon estimate that the Taiwanese spread two viruses, known as the Bloody 6/4 and Michelangelo, damaging 360,000 computers in China at a cost of $120 million. The Crypt Newsletter wrote that, contrary to such claims, neither virus is spread on the internet or by networked computer connections, and that both have been essentially distinct on Western computers. This has resulted in a controversy over whether media reporting is accurate, since it takes a modicum of technical expertise to report on such incidents. Controversy also persists over the role of vested interests in exaggerating cyberwar reports to receive funding – for instance the FBI receives special funding for such projects outside normal budgets (Sun.soci.niu.edu/~crypt).

Another example was the 'Fluffy Bunny', which for a six-month period starting in mid-2001, penetrated the networks of several top internet firms, including Exodus, VA Software and Akamai. The hackers also vandalized websites operated by leading security outfits, including the SANS institute (McWilliams 29 July 2002). However, despite the thousands of hackable holes found in email, websites, files and operating systems, few of the hacks in 2002 turned out to have any real impact on most computer users. The Klez virus infected some machines and the Linux Slapper worm made more work for system administrators for while. Sweeny, an internet security expert, commented in December 2002 that 'the average user wouldn't know a hack if it walked up and hit them. And many of the so-called security holes require a very specific event to occur and the odds are very slim that it will occur' (Delio 30 December 2002). This is an opinion also shared by Jim Lewis, a 16-year veteran of the State and Commerce Departments, who compiled an analysis for the Centre for Strategic and International Studies. 'A hacker or a large group of hackers would need to find vulnerabilities in multiple systems to significantly disrupt the power supply and even then, an attack might only disrupt service for a few hours' (Shachtman 20 December 2002).

Internet security analysis

There are two components in computer security: virus protection (lack of trust in outside sources) and authentication (verifying that a person is known to the user). But the first notion of 'common sense' is far less effective in securing computers. Users expect their machines to do as they are told, but such expectations leave them prey to low-level, but insidious, information-warfare attacks (Libicki and Shapiro 2001: 441). In Britain, three-quarters of all companies have been hit by hackers. Each attack costs £120,000 on average. By contrast, it is more approachable to question what form an attack takes than to question who the attacker is. It is tempting to see intrusions in terms of a pyramid that goes from transient vulnerability probing and defacing of websites at the base, to large-scale efforts to undermine the critical missions of an organization or the critical functions of a nation at the top – and to suggest that there is an inverse relationship between frequency and significance, with many trivial incidents and comparatively few of the more serious incidents. This lack of congruence between limited intent and far-reaching consequences stems from the capacity of worms and viruses for infinite replication and multiplication.

Consider, for example, the 'I love you' virus, which caused an estimated $6.7 billion in damages in its first five days. It was caused by a single individual with poor support and little preparation. As Shimeall *et al.* argue, the lesson was very clear: 'the development of national and global information systems has outpaced appropriate safeguards and security measures. This provides new targets and new opportunities for criminal organizations, terrorist groups and hostile nations' (2001: 2).

One further barrier here may be the structure of the intelligence community itself, which, for constitutional reasons, has been partitioned into domestic and foreign intelligence. While this distinction is critical in the physical world, it vanishes rapidly in the cyber-world. The intelligence community has been left with the uncomfortable choice of violating important barriers, an unacceptable option, or yielding the cyber-world to transnational threats, also an unacceptable option (Shimeall *et al.* 2001: 10).

In Shimeall *et al.*'s internet security analysis, six problems are prominent: i) the lack of borders, which differs from the more familiar domains of intelligence; ii) the identification of intruders, which is critical to the assessment of the challenge and the nature of the response; iii) the forms of attack, which either fall far short of what was intended, or far exceed what the perpetrator initially envisaged; iv) the question of how intrusions take place, which is the most technical aspect of the problem and the easiest to answer; v) the legal issue of who gets notified when analysts attribute an intrusion to a foreign individual; vi) the structure of the intelligence community itself, which for constitutional reasons has been partitioned into domestic and foreign intelligence (2001: 5–10).

Another problem in internet security analysis is that of disclosing computer intrusions publicly. US cybersecurity director Richard Clark, in November 2002, and virtually all software companies insisted that software vendors should have a chance to fix problems before security researchers disclose them publicly (Delio 19 November 2002). Researchers have criticized Symantec for encouraging people to create and release malicious codes. 'Given that Symantec also sells security and antivirus software, I think there is a terrible conflict of interest here' (Delio 19 November 2002). Moreover, security experts caution IT managers not to publicize security holes simply to embarrass vendors into action, because within hours, black-hat hackers would know the vulnerability and use it. One test case is NASA, where more than 24 per cent of attempted hack attacks penetrated the system in the third quarter of 2000; but by the end of 2001, after scanning and fixing system flaws, only a fraction of 1 per cent of attacks were successful.

More specifically, methods of intrusion are the online equivalent of military tactics and, just as in the military world, there has historically been a dialectic between defence and offence. Similarly, on the internet, there is a dialectic between protection and intrusion. Intruders manipulate the sources of intrusion and the online records of activity. The sources of intrusion are manipulated either by staging intrusions through a series of already-intruded and corrupted hosts, or by falsifying source information from the network traffic. Online records on the other hand are falsified either by directly modifying the records themselves or by replacing the monitoring software that produces these records (Shimeall *et al.* 2001: 8). The following are of paramount importance: details of the victim's infrastructure, the nature of the intrusion, identity clues left by the intruder, network traffic flow and intrusion tools left as traffic artifacts on the victim hosts.

The latest in internet security is that angry computer users fed up with attacks from internet hackers are hacking back. They are buying counterstrike software to turn the tables on the hackers. The new software could spawn a generation of vigilante net users prepared to take the law into their own hands to protect their hard drives (Harding 10 June 2004). In other words, the users of the internet are not easily controlled or subverted. As Foucault suggested, every exercise of power – in our case whether by a government or a cyberwarrior – will create a resistance somewhere, even among the 'public'.

Sociopolitical implications

To continue, Gombert believes that competitiveness in information technology depends on economic and political freedom and on integration into the core. By integration into the core, he means the core of power in Western democratic states. The idea behind this is that the constantly changing nature of this techno-logy favours open economic systems:

> The main economic uses to which information technology is put – distributing information, decentralising functions and decision-making, cre-ating horizontal links, improving producer-consumer contact, sharpening external awareness and adaptability – correspond with strong market forces. Even if the supply of information technology becomes less dependent on economic freedom over time, the demand will not. Therefore, we should expect capitalist systems to retain their advantage through the information age.
>
> (Gombert 2001: 54)

What this amounts to is that other powers that want to get into the game where open market democracies dominate will have to 'open up to the pressures for reform and freedom that create modern knowledge-based power' (Gombert 2001: 61). According to Gombert there are three reasons why this is the case. First, competitiveness in information technology depends on economic and political freedom and on integration into the core. Second, military power and other forms of national power depend on broad-based competitiveness in the creation and use of information technology. Third, integration into the core creates shared stakes that eclipse, or at least qualify, power politics and which point towards a democratic commonwealth of interests and values. Interestingly, Gombert argues:

> Direct support for dissidents or embryonic democratic institutions is increasingly available both from the governments and nongovernmental organizations of the democratic core, thanks to (what else?) information technology. The penetrability of even self-isolated societies is growing, especially when sophisticated transnational 'civil society' groups make it their business to network with the oppressed.
>
> (2001: 50)

Most information products and services work well only when embedded in a society whose skills and infrastructure are undergoing a larger information revolution. These technologies are increasingly interdependent, especially as computer networking expands; individual items of hardware are of limited utility. What good are desktop computers without networks and a steady diet of software upgrades? Information technology is constantly being modified, enhanced and overtaken by better ideas, leaving importing states to engage in an expensive and never-ending game of catch-up technologies which have been conducive to state power, even to coercive state power.

On the socio-political aspect of cyberconflict, the lack of central control, cheap remote access from anywhere and interconnectivity promise to render cyberspace an unprecedented power-enhancer against repressive regimes and societies. This is one of the reasons that the US has devoted so many committees to monitoring the web and defending electronic frontiers. In a report entitled 'Strategic Assessment: the Internet', dating back to July 1995, Charles Swett, assistant for Strategic Assessment, office of the Assistant Secretary of Defense for Special Operations and low intensity conflict, writes:

> Current information about conflicts placed on the Internet in real time by on-the-scene observers and alternative news sources will be voraciously devoured by the world audience and will have an immediate and tangible impact on the course of events. Video footage of military operations will be captured by inexpensive, hand-held digital video cameras operated by local individuals, transformed unedited into data files, and then uploaded into the global information flow, reaching millions of people in a matter of minutes. Public opinion and calls for action (or calls to terminate actions) may be formed before national leaders have a chance to develop positions or to react to developments.
>
> (July 1995)

This is more or less what happened with the anti-globalization movement, with the anti-war movement and in various other conflicts. Indicative examples are the photos of naked prisoners at Abu Ghraib, as well as the video of Nick Berg's murder by fundamentalists in Iraq shown globally through the internet. Swett makes three other points which are relevant here. Firstly, if protest groups opposing oppressive regimes persuade trade unions or organizations from richer countries to equip them with the necessary facilities, they can reach a world-wide audience, a situation that any government would find difficult to control. Swett's claim is supported by the efforts of several dissidents, as the Zapatista, East Timorese, Tunisian, Chinese, Iraqi, African and Indian social movements have demonstrated. Secondly, such a threat would be very difficult to counter, because personal contact of people living under oppressive regimes with the free world would 'open' their eyes to the distorted propaganda of the regime and dis-credit their government (Swett July 1995). Third, the internet has the potential to significantly influence the course of a conflict. In Swett's words:

Thus the Internet can indirectly play an important role in the way the world deals with a conflict, without having substantial physical presence within the conflict. The Internet can play an important positive role during future international crises and conflicts. In the chaotic conditions usually present in such situations, normal government and commercial reporting channels are often unreliable or unavailable, and the Internet might be one of the few means of communication present.

(July 1995)

To continue, the largest and most active international political groups using the internet appear to be the San Francisco-based Institute for Global Communications (IGC) and the Association for Progressive Communications (APC). Therefore, a review of the IGC can provide a good perspective on the breadth of US Department of Defense (DoD)-relevant information available on the internet. According to a text file placed on the IGC's publicly accessible internet site:

IGC provides computer networking tools for international communication and information exchange. IGC is the U.S. member of the Association for Progressive Communication (APC), a coalition of computer networks providing services to over 25,000 activists and organizations in more than 130 countries. The IGC networks – PeaceNet, EcoNet, ConflictNet and Labor-Net – together with APC partner networks, comprise the world's only computer communications system dedicated solely to environmental preservation, peace, and human rights. New technologies are helping these worldwide communities cooperate more effectively and efficiently ... send and receive private messages to and from more than 18,000 international peace, environmental and conflict resolution users on our affiliated networks or to millions of users on virtually any other network ...

(Swett July 1995)

The Association for Progressive Communication (APC) provides low-cost computer communication services and information-sharing tools to individuals and NGOs working on social issues. So, in a way, we find 'solidarity in cyberspace' (Mansbach 2000), where a new 'electronic fabric of struggle' is being constructed, helping to unite activist movements around the world (Cleaver 1999). However, as Arquilla and Ronfeldt (1998: 71) argue, there is concern about the internet being used for crying wolf and for manipulation by people with hidden agendas. The concern is that government actors may post misinformation and disinformation on the internet in order to provoke an overreaction that embarrasses the activists. The internet is a potent tool for inflammatory rumour, as well as 'black' and 'grey' propaganda, in that actual affiliation of the provider of information can be masked easily and any news materials that are put on the web can be transformed so as to make faked events appear true (Hosmer 1999).

Further, in recent conflicts the internet has been cited by both parties as playing a key role. As early as 1989, during the heyday of the pro-democracy movement in China, Chinese students used the internet to exchange information, and news was faxed to China to inform people who were unable to obtain information because of censorship imposed by the Chinese government. In October 1989, after the Tiananmen massacre, students organized a rally in Washington through email, with 40,000 participants across the country. Also, they successfully lobbied Congress, using email as the major communication medium to pass the Chinese Students Protection Act of 1992, which protected Chinese students from being forced to go back to China and face prosecution (Wang undated).

Also, new global conflicts have involved cross-fertilization and combination of energies generated at a local, grass-roots level. As Cleaver points out,

> local conflicts between citizen groups and governments have expanded into global efforts in response to two things: first, to a spreading uniformity of policies and international agreements among governments to implement world-wide sets of rules and second, to the resultant perception of common interests in challenging not only those rules but any set of uniform mandates unrelated to local situations.
>
> (1999)

An example mentioned by Cleaver is when OECD opened to dialogue after failing to pass the Multilateral Agreement on Investment, which aimed at a global set of rights for corporate investors. This temporary defeat has been credited to the use of the internet in organizing a campaign against MAI in countries around the world (Cleaver 1999). This pressure caused the OECD to halt negotiations for six months in April 1998, and France pulled out of the negotiations, undermining them.

Nevertheless, there have been criticisms of activists using the internet. The first criticism is that on certain occasions the target is badly chosen because they are not properly connected to the group their actions are supposedly supporting. Second, the use of hacking tactics denies the opponent of freedom of speech and provides a precedent for activists to be attacked in the same way. Third, there is the problem of gaining legitimacy through providing evidence that the action involved a fair number of people and not just a small group of hackers (Cleaver: 17). Moreover, cyberspace holds information about various types of political struggle which have not yet been connected, because the availability of information does not guarantee that the necessary connection will be made to generate action. Worse, the very amount of information and the number of groups using it will make this situation deteriorate.

However, unlike newspapers, radio and television media where feedback is slow, the internet enables quick interaction, and email lists with ongoing flows of discussion form an alternative community of debate, outside traditional policy-making institutions (Cleaver 1999). Moreover, the groups involved have characteristics which reflect the democratic character of the medium. Users of the internet are colour-, class- and gender-blind and, in contrast with face-to-face interaction where the status of the speaker is primary, attention is focused on what is said, not who is saying it (Wang undated, citing Craut 1991).

Even though the internet's potential for mobilizing world opinion is likely to be diluted by the increasingly large number of issue-oriented groups contending for attention in cyberspace, it will still remain a potent instrument for rallying like-minded people to support or oppose particular actions and causes (Hosmer). Unfortunately, news materials that are put on the web can be masked easily and any visual 'news' materials can be transformed to make faked events appear true. Moreoever, new problems constantly arise, such as spamming, which has been tormenting users for some time now. British lawmakers plan to use a new tactic to stop the torrent of junk email spam that floods in from overseas, extraditing the mass-mailers and bringing them to trial in the United Kingdom. While the initial policy would be used to target spammers, it could be expanded to include suspects in other cybercrime offences such as virus-writing and hacking (Reuters 30 October 2003).

Conclusion

This chapter has examined the environment of cyberconflict by defining the terms involved and the approach to be taken, looking at cyberterrorism, information warfare, internet security analysis and the sociopolitical implications of cyberconflict. The central argument of the thesis is that two types of cyberconflict occur: firstly, between ethnic or religious groups fighting in cyberspace as they do in real life; and secondly, between a social movement and its antagonistic institution (hacktivism). This distinction is needed so that social movement theory can be used to explain sociopolitical cyberconflicts, and so that conflict theory can be used to understand ethnoreligious cyberconflict. These topics will fall under the more general umbrella of media theory, because we are discussing a new medium.

These issues were addressed by analysing the internet not in terms of a mass medium in the traditional sense (i.e. its influence on the political outcome of a conflict) but, rather, as a resource used by the opposing parties in a conflict. In this context, a discussion Wolfsfeld's political contest model emerged, as did discussions of Deleuze and Guattari's rhizomatic politics and Arquilla and Ronfeldt's argument that the information revolution is altering the nature of conflict by strengthening network forms of organization over hierarchical forms.

Following these theoretical insights, an analysis of the hacking phenomenon was provided. The motivations for hacking, as explained by Taylor, include feelings of addiction, the urge of curiosity, boredom with the educational system, peer recognition and political action. Hacking can potentially perform a variety of benevolent services to the security industry – for instance, as a watchdog counter-response to the use of surveillance technology and data gathering by the state, contributing to a general climate of improved security consciousness.

The section on information warfare drew attention to the idea that territory, population and natural resources are becoming less important relative to human capital and the possession of information. Moreover, intelligence officials are trying to move away from their old hierarchies and towards a structure in which every soldier, every drone and every general are connected by computer networks. This is because power seems to be migrating to non-state actors, who are more flexible and responsive than hierarchies at using information to improve decision-making.

Despite this trend, doubts have been raised as to the real or exaggerated nature of cyberwar. Controversy persists over whether media reporting on cyberattacks is accurate, and whether there is a vested interest in exaggerating cyberwar reports to receive funding. Nevertheless, the US government has created dozens of committees to deal with information warfare. Notably, the full extent of the US cyberarsenal is among the most tightly held national security secrets, even more guarded than nuclear capabilities. Moreover, the Bush administration has toughened anti-hacker laws since September 11 and increasingly lobbies governments to cooperate in international computer crime investigations.

The section on internet security analysis looked at incidents like the Slammer, Blaster and MyDoom viruses which made the year 2003 the worst in computer-virus history, leading to the conclusion that the development of national and global information systems has outpaced appropriate safeguards and security measures. This provides new targets and new opportunities for criminal organizations, terrorist groups and hostile nations. Another problem in internet security analysis is that of disclosing computer intrusions publicly.

On the sociopolitical implications of cyberconflict, which ultimately justify its study, it was found that if protest groups opposing oppressive regimes persuade supporters in richer countries to equip them with the necessary facilities, they can reach a worldwide audience, a situation any government would find difficult to control. Also, in international crises and conflicts where normal government and commercial reporting are unreliable, the internet might be one of the few means of communication present. An example of alternative reporting is the internet's effect on the 2003 Iraq conflict, discussed in the section 'The internet's effect on media coverage' (pages 178–187) in Chapter 6. On the downside, there is information about all types of political struggle in cyberspace, which have not yet been connected, because the availability of information does not guarantee that the connection needed to generate action will be made.

In the following chapter elements of social movement theory will be used to examine sociopolitical cyberconflicts, in which social movements use the internet against antagonistic institutions, and also address the issue of dissidents using the internet to oppose their respective governments.

4 Sociopolitical cyberconflicts

This chapter looks at sociopolitical cyberconflicts, providing an explanation of the forms and types of this kind of conflict and discussing instances of it, such as the anti-globalization movement. It also includes an extensive examination of Chinese cyberdissidents and the Chinese government's cyberstrategies, as well as a brief look at internet censorship and dissidents internationally. The main argument of the chapter is that groups involved in sociopolitical cyberconflicts use the internet as an organizational and mobilizational resource, attempting to reframe issues and take advantage of the openings of the political opportunity structures offered by new communication technologies. Lastly, the examples will be explicitly linked with the proposed integrated theoretical framework.

Sociopolitical cyberconflicts

Sociopolitical cyberconflicts can be placed into two broad groups. First, there are those primarily concerned with global issues such as the environment, issues in relation to which the level of negotiation with governments is open for debate. Second, there are groups much less inclined to negotiate with governments, as they are concerned with such issues as their own liberation from the control of the state (Arquilla and Ronfeldt 2001: 15). One aspect of the problem seems to be that, with the explosion of the size of the internet, protests and political activism have entered a new realm. Political activism on the internet has generated a wide range of activity such as using email and websites to organize web defacements and denial-of-service attacks, as described above. These politically motivated attacks are called hacktivism. Stanton McCandlish, program director of the Electronic Frontier Foundation, describes it as follows: 'a kind of electronic civil disobedience in which activists take direct action by breaking into or protesting with government or corporate computer systems. It's a kind of low-level information warfare, and it is on the rise' (in Phrack Magazine December 1998). The main purpose of these activists is to influence or challenge public or enemy opinion and battle for media access and coverage. In many respects, then, the archetypal netwar design corresponds to what earlier analysts called a 'segmented, polycentric, ideologically integrated network' (SPIN):

By segmentary I mean that it is cellular, composed of many different groups ... By polycentric I mean that it has many different leaders or centers of direction ... By networked I mean that the segments and the leaders are integrated into reticulated systems or networks through various structural, personal, and ideological ties. Networks are usually unbounded and expanding ... This acronym [SPIN] helps us picture this organization as a fluid, dynamic, expanding one, spinning out into mainstream society.[1]

The quality of networks contrasts sharply with the tendency to forge social and political order through mutual identifications with leaders, ideologies and memberships in conventional social and political groups. Gerlach applied the SPIN model to contemporary global protest networks. First, he replaced the idea of polycephalous organization with polycentric order, indicating that, like earlier SPIN movements, global activist networks have many centres or hubs but, unlike their predecessors, these hubs are less likely to be defined around prominent leaders. In addition, he noted that the primary basis of movement integration and growth has shifted from ideology to more personal and fluid forms of association (Bennett 2004: 126).

Sociopolitical cyberconflict could be seen as taking two forms, one being when proper hackers attack virtually chosen political targets, the other being when persons organize through the internet to protest, or carry through email a political message. One of the first examples of hacktivism conducted by hackers was back in 1999 when the 24-member Legions of the Underground called for an information warfare campaign against human rights violations (Glave 8 January 1999). However, hackers around the world condemned the initiative by issuing a statement signed by well-known names in the computer underground such as 2600, the Chaos Computer Club, L0pht, Heavy Industries, Phrack, Toxyn, Cult of the Dead Cow, !Hispahack, Pulhas and several Dutch hackers. The joint statement said: 'Legions of the Underground will do little to alter existing conditions and much to endanger the rights of hackers around the world. Declaring "war" against the country is the most irresponsible thing a hacker group could do. This has nothing to do with hacktivism or hacker ethics and is nothing a hacker could be proud of' (Glave 8 January 1999).

One kind of online activism is the distribution of email petitions, where each person adds his or her name to a list of names that will then be forwarded to friends and acquaintances. The original email petition then splits into several lists. As a result, when the originator of the petition tries to count the number of people who signed the original list, he or she would have to sift through potentially thousands of duplicate signatures. Also, the opportunity for forgery looms large. For these reasons, politicians treat them as dubious. On the other hand, petitions posted on a website show the signer actively logging onto the site and allow the reader more time to engage with the issue. Web-based petitions are utilized more in the activist world, but email petitions contribute to electronic activism in their capacity for getting the word out of dates and times of organized protests, demonstrations and coordinated activities (Lebowitz 22 May 2003).

Cronauer has looked at how the use of electronic mailing lists contributed or hindered mobilization of list subscribers, by examining messages sent through two electronic mailing lists. Specifically, she has examined how groups form their goals and activities; how individuals respond to these online framing efforts; how structural features of electronic mailing lists shape online messages; and how the context such lists are used affect online interactions (Cronauer November 2004). Cronauer's two electronic mailing lists, in the course of several months, were used to organize opposition and create alternatives to the globalization agenda pursued and represented through an APEC summit in Canada, and through both a European Union and a G7/G8 summit in Germany. In both cases, local groups formed specifically to protest the respective meetings to exist only temporarily. The mailing lists were, however, also used by other groups, including those the newly formed groups came out of, thereby allowing for long-term involvement. She identifies the following aspects of online mobilization: outreach, networking, increase of visibility, opinion formation and strategic planning. Even more importantly, she found that electronic mailing lists impact social activist groups in various ways:

> they may result in different (be that additional or alternative) means of communication and hence in different ways to communicate; inhibit individuals from becoming mobilized that otherwise may have become mobilized; or prompt individuals to become mobilized that otherwise may not have become mobilized.
>
> (Cronauer November 2004)

An example of an organized protest through the internet on a small scale occurred in Madrid in November 2002, when the cyberpunks, who gather at the site Ciberpunk, discovered that Gallardon, president of the Community of Madrid, had proposed a law that would classify Madrid's cybercafés as casinos. If the law had passed, minors would not have been allowed in cybercafés, from which roughly 20 per cent of Spain's young internet users connect to the internet. In October, the country passed a law requiring all internet users to register with the government. But Gallardon announced he would not support the law, an action reported in Spanish newspapers as a win for cyberpunks, who credit their victory to the internet and getting non-net users involved. 'The Internet is still a strange land to our politicians, who are not used to facing such an immediate reaction', said activist David de Ugarte (in Delio 26 November 2002).

Tim Jordan writes that social activists or hacktivists have found two uses for the internet, Mass Virtual Direct Action (MVDA) and Individual Virtual Direct Action (IVDA) (2001: 8). According to Jordan, MVDA involves the simultaneous use by many people of the internet to create electronic civil disobedience. An example of Mass Virtual Direct Action on the internet is the one organized by the Electronic Disturbance Theatre (EDT). To demonstrate solidarity with the Zapatistas, an estimated 10,000 people from all over the world participated in the sit-in on 9 September 1998 against the sites of President Zedillo, the

Pentagon and the Frankfurt stock exchange, delivering 600,000 hits per minute to each. There are two characteristics of this type of hacktivism. First, hacktions are not aimed at halting a target permanently, but have symbolic dimensions. Second, MVDA activists rarely try to hide their identities, and seek public debate and discussion. IVDA is different from MVDA in that it could be taken by an individual and does not depend on a mass protest. The actions taken are either semiotic attacks (i.e. defacements), computer intrusion or attacks on network security.

According to more recent analysis from Jordan and Taylor, hacktivism breaks down into two broader streams of action:

1 Mass virtual direct actions, which use cyberspatial technologies of limited potential in order to re-embody virtual actions.
2 Digitally correct actions, which defend and extend the peculiar powers cyberspace creates.

(2004: 116)

As Taylor and Jordan explain, mass action hacktivism is the kind of action most closely associated with the anti-globalization movement. Here we find the most direct attempts to 'turn' traditional forms of radical protest, such as street demonstrations, into forms of cyberspatial protest. Digitally correct hacktivism, consists of online direct actions that are influenced perhaps more by the history and technical concerns of hacking than by the more directly political concerns of anti-globalization protests (2004: 68). Mass electronic disobedience creates the complex situation in which an embodied presence at a terminal uses direct action to become an abstract virtual presence, which in turn joins with other abstractions to jam up a virtual site. Legitimation comes from the embodied presence, action from the virtual presence (Taylor and Jordan 2004: 170).

An example of mass action hacktivism would be the Seattle anti-World Trade Organization protests at the end of November 1999, which were the first to take full advantage of the alternative media network via the internet. Protestors used mobile phones, direct transmissions from independent media feeding directly onto the internet, personal computers with wireless modems broadcasting live video and a variety of other network communications. As Naomi Klein pointed out in the *New York Times*, this was a movement

born of the anarchic pathways of the Internet ... the most internationally minded, globally linked movement the world has ever seen ... When protesters shout about the evils of globalization, most are not calling for a return to narrow nationalism, but for the borders of globalization to be expanded, for trade to be linked to democratic reform, higher wages, labor rights and environmental protection. This is what sets the young protesters in Seattle apart from their 60s predecessors.

(in Kneen 3 October 2002)

During the anti-WTO protests, hacktivists managed to acquire the URL www.gatt.org, using the GATT address for a parody WTO site, looking identical to the original WTO one, but included a text criticizing WTO trade policies. Another famous hacktion was an MVDA by the Electrohippies, which included a virtual sit-in with a downloadable web page aiming to flood the WTO server. The Electrohippies claim that 400,000 hits in this MVDA had slowed down the WTO and at times completely halted it (Jordan 2001: 8). The Electrohippies developed what they called a client-side DDOS, whose central feature was that each client or end computer had to choose to initiate the attack on its own, thereby depending on many people to initiate a mass action (Taylor and Jordan 2004: 77). Electrohippies Collective 2000 stated that 'we have to treat cyberspace as if it were another part of society. Therefore, we must find mechanisms for lobbying and protest in cyberspace to complement those normally used in real life' (in Taylor and Jordan 2004: 78).

The anti-WTO protesters were able to initiate a newsworthy event, putting the other side on the defensive. Using the internet, they could send stories directly from the street for the whole world to see, rendering the flow of information uncontrollable. They were able to mobilize support by promoting an alternative frame for the event. More recently, in September 2003, during the WTO meeting in Cancun, activists used a peer-to-peer video-sharing service called v2v to transmit broadcast-quality video of the protests to television stations and other activists. They also set up wireless networks at the protest welcome centre in a nearby town, streaming recordings of interviews and speeches over the internet for rebroadcast on participating radio stations. Two tactics were not used in Cancun: the use of Distributed Denial of Service attacks to shut down the WTO site; and the 'flash mob' concept, a way of mobilizing large groups of people by sending out a single message to awaiting participants. With just 10 per cent of Mexico's population online, it evidently wouldn't work (Arasavala 28 August 2003).

There is a debate between hacktivists concerning Denial of Service attacks and web defacements. On the one hand, there are activists who claim that such actions run contrary to other people's right to freedom of speech and, on the other, there are those who view these actions as the only way to get the public's attention. In any case, the fact is that web defacements cannot be dismissed as electronic graffiti and Denial of Service attacks as nuisances, because there is concern by online companies that it could affect share prices, reduce earnings and cause damage to reputation and customer confidence. Indeed, this is the reason why there should be more analysis of the reasons underpinning hacktivism and its political rationale.

Real hacktivism is not supposed to be about taking down servers and websites and grafitting webpages. Hacktivism, as defined by the Cult of the Dead Cow, the group of hackers and artists who coined the phrase, was intended to refer to the development and use of technology to foster human rights and the open exchange of information (Delio 14 July 2004). Professor Ronald Deibert from the University of Toronto's Citizen Lab, which sponsors and develops technology used by activists, said real hacktivism is fast becoming understood and accepted by mainstream human rights activists, and is now being supported by large foundations like the Soros Foundation, Markle Foundation and Ford Foundation, which funds groups such as Privaterra, eRiders and Indymedia, which use technology to defend civil rights (Delio 14 July 2004). Additionally, a network of radical internet service providers has sprung up, including Riseup, Mutualaid, resist.ca, Interactivist, OAT and others. Radical geeks brought together by anti-globalization protests and the Indymedia network have developed their own international network of mutual aid, support, skills sharing, free software and solidarity (Munson 30 November 2004).

A very explicit example of the schism in the hacktivist community between those that support hack attacks and those who view them as an infringement of free speech occurred during the Republican National Convention (30 August to 2 September 2004). Protest organizers used websites and printed fliers to direct thousands of supporters to rallies and marches. But police also were surfing the web and collecting those same fliers in order to mobilize their own sizable forces. In many cases, police arrived at protest sites well before the demonstrators, pre-empting marches and rallies before they coalesced, using the protesters' own lines of communication to keep most of them well away from Madison Square Garden, the convention site (Zucchino 4 September 2004). *The Times* reported that during the weeklong protests, the police were monitoring websites. Authorities at the highest level of government seemed to be paying attention. Just as the week of protests was kicking off, the Department of Justice announced it had opened a criminal investigation into the New York City Indymedia centre for posting names of Republican delegates (Scahill 9 September 2004).

In addition to the various groups using SMS text messaging to send out action alerts, warnings, news and announcements, the New York Independent Media Center (IMC) set up an automated information line that activists could call 24 hours a day to hear breaking news from Indymedia and to listen to a live streaming broadcast from the A-noise radio collective, which was broadcasting live reports from the streets. The SMS messages alerted activists of routes that remained open to travel to protests when police blocked off large sections of the city. It alerted Indymedia journalists of where cameras were needed to document protests, legal observers of real-time rights violations and activist medic teams of where people were in need of medical attention (Scahill 9 September 2004). In contrast to previous protests, Indymedia was not only available to people at home but went mobile. Evan Henshlaw-Path, the Indymedia tech activist who developed the info-line concept, commented:

> Our task is to help facilitate horizontal communication and information dis-tribution to all activists in the streets. The police want to keep the protests under control and stay a step ahead of the protesters. So, all of this commu-nications infrastructure helps on a tactical level. We've appropriated techno-logy as an essential tool for radical social change.

> (Scahill 9 September 2004)

Nevertheless, the internet was used not only as a resource but also as a weapon, as hacktivists launched a campaign of electronic disobedience to coincide with the demonstrations against the Republican National Convention.[2] They targeted major credit card corporations, disrupting various right-wing fascist groups, and performed an electronic sit-in against Republican websites. Credit card numbers stolen from major news corporations were used by anti-RNC hacktivists to make $2,600 in donations to various humanitarian and civil rights organizations. A right-wing fascist organization known as ProtestWarrior was also hacked and defaced a week before the convention began: 'By infiltrating and crashing legal, peaceful assemblies, the Protest-Warriors are fighting against the democratic process while claiming to uphold the core values of this country' (Indymedia 5 September 2004). The mobile/home phone numbers, names, addresses and passwords of the site's lead organizers along with the email addresses of all ProtestWarrior members were posted to the site and emailed out to every member (Indymedia 5 September 2004).

The Internet Liberation Front hacked and defaced six Republican websites. These hackers do not believe hacktivism to be a violent tactic. By using direct action techniques such as website defacements and financial disruption, 'hacktivists' believe they are putting pressure on politicians to make progressive and revolutionary changes (e-resistance@yahoogroups 'Hacktivists hack Republicans' 18 January 2005):

> The will of the people was not expressed in these elections. Imperialist war, tax cuts for the rich, and ecological destruction are not in the interest of working people or the stability of our global society. The Bush administration are rich lying thieves, these inaugurations are a joke, and the whole system is corrupt.

Speaking of protests, on the down side, the epic qualities of the best demonstrations, both in terms of size and drama, are lost in cyberspace. While mass action hacktivism clearly involves symbolic actions, and it can draw large numbers of people together to protest, some of the qualities of a symbolic demonstration are lost (Taylor and Jordan 2004: 80). However, as stated earlier, 'online protests have the advantage of being able to pass a great amount of information, in forms that mean people can take a little and explore a lot. In non-virtual spaces it is difficult to hand over more than a leaflet, even if volumes could be devoted on the particular cause' (Taylor and Jordan 2004: 80). As these researchers argue, in the overarching context of virtual times, and born both from hacking and from the anti-globalization movement of the twenty-first century, hacktivism is perhaps the first widespread social and political movement of the new millennium (Taylor and Jordan 2004: 43).

Chinese dissidents

> The stronger the state, the weaker its encouragement of institutional partici-
> pation and the greater the incentive to confrontation and violence when
> collective action does break out.
>
> (Tarrow 1996: 46)

The argument for state strength runs like this: centralized states attract collective actors to the summit of the political system, whereas decentralized states provide a multitude of targets at the base (Tarrow 1998: 81). Gary Marx divided repressive actions into different categories according to their specific aims: the creation of an unfavourable public image, disinformation, restricting a movement's resources and limiting its facilities, derecruitment of activists, destroying leaders, fuelling internal conflicts, encouraging conflicts between groups and sabotaging particular actions (Della Porta 1996: 65).

Direct support for dissidents or embryonic democratic institutions is increasingly available both from the governments and non-governmental organizations of the democratic core. The penetrability of even self-isolated societies is growing, especially when sophisticated transnational 'civil society' groups make it their business to network with the oppressed. Laclau and Mouffe offer a useful way of thinking about discourse and ideology, especially since it neither restricts methodological enquiry, nor necessarily follows their own logic. Discourse is 'the structured totality resulting from ... practice', when elements are those differences that are 'not discursively articulated' because of the floating character they acquire in periods of social crisis and dislocation (Howarth *et al.* 2000: 7).

In a RAND report on the political use of the internet, Chase and Mulveron describe and analyse Chinese dissidents' use of the internet and the Chinese government's counter-strategies (Chase and Mulveron 2002). The dissidents have used email spamming, set up proxy servers to access blocked sites, set up sophisticated websites, used email lists, bulletin board sites, file trading and e-magazines and a particular group even organized a mass demonstration and a press conference through the internet. Nevertheless, there have been arguments that China will not change politically until the Chinese people demand change. The internet may help citizens of the People's Republic break free of centuries of kowtowing to imperial rule, but it is not going to cause that rupture by itself (Leonard 21 March 2002).

This report is important for understanding how the internet could be used to achieve political aims, because both the dissidents and the Chinese leadership itself have realized the political potential of the internet. Moreover, as the RAND report indicates, both groups are already engaging in an online battle over the internet, which touches reality as well.

The question the two authors address is whether the internet provides dissidents with new tools to break through the barriers of censorship and ultimately undermine the power of non-democratic regimes, or whether it is more likely that those authoritarian governments will use the internet as another instrument to repress dissent, silence their critics and strengthen their own power (Chase and Mulveron 2002: xi). The main dissident groups examined in the report are the Falungong religious sect, the Chinese Democratic Party (CDP) and the Tibetan community-in-exile.

The Falungong is a religious sect banned in China, whose membership is estimated to be in the millions. It combines meditation with certain quasi-spiritual beliefs. In August 2002 the world's leading psychiatric association decided to look into reports that China is silencing political dissidents by confining them to mental wards, where some including members of the Falungong sect are drugged or undergo electric shocks. The group said that among those are nearly 500 members of the Falungong. Thousands of its followers have reportedly been arrested and sent to labour camps (Greimel 26 August 2002). On 20 July 1999, hundreds of key members were arrested in the middle of the night, and the next day the sect was banned for allegedly spreading 'superstitious, evil thinking'. In October 1999, Falungong held a clandestine conference in Beijing to tell the world about police beatings of detained members. This conference was attended by foreign journalists and, again, was organized through the internet (Chase and Mulveron 2002: 10).

Because the sect is banned in China, email is vital for communication and information dissemination. The sect's leader Li Hongzhi, now living in the US, has set up email lists to connect followers in the US and China. After his retreat from 'public' life, all his new writings are disseminated through email lists via Minghui Net. Email lists gained even more importance after the Falungong bulletin boards and chatrooms were attacked by the state (Chase and Mulveron 2002: 17). The official Chinese news agency, Xinhua, stated in 19 February 2004 that a court in western Chongqing had found three men and two women guilty of 'vilifying the government's image through spreading fabricated stories on persecution of cult practitioners', and had given them prison terms of five to 14 years. Reporters Without Borders has called for the release of 22 Falungong members imprisoned for publishing news on the internet about the spiritual movement. 'The crackdown on members of this spiritual movement is completely unjustified. The five Internet users were convicted for posting online what is already very well-known to human rights organizations, that members of Falungong are systematically tortured in prison', commented Julien Pain, who researches internet speech issues for the group (Wired.com 24 February 2004).

The Chinese Democratic Party announced its existence on 24 June 1998. Their formation was timed to coincide with Clinton's visit to China. However, as US–Chinese relations deteriorated, Beijing decided on a crackdown against the CDP and, in late December 1998, the CDP's most prominent members, Wang Youcai, Xu Wenli and Qin Yongmin, were sentenced to long prison terms on charges of endangering state security. At least two dozen more CDP members have since been imprisoned, and many others are being held in detention. Despite the crackdown, a number of CDP members remain active on the mainland. The CDP has also developed an organization-in-exile in the United States (Chase and Mulveron 2002: 12–13). Several CDP members assert that the use of email and the internet was critical to the formation of the party and allowed its membership to expand from about 12 activists in one region to more than 200 in provinces and municipalities throughout China in only four months (Chase and Mulveron 2002: 15). The Chinese Democratic Party is an example of a sociopolitical group seeking participation, power and democracy using information technology techniques to voice those demands, in a country where power and political participation are exclusive rights of the communist party, and where dissidents are arrested and democracy is non-existent.

The third dissident group examined is the Tibetan government-in-exile and its supporters, and more than a dozen associated NGOs. Two-way email communication is an important channel for Falungong members, Tibetan exiles, and various NGOs and human rights advocacy groups to pressure US government officials. Nevertheless, one official Chase and Malveron interviewed said that, because of the vast quantity of email received on a daily basis, the messages have largely lost their effect (Chase and Mulveron 2002: 18).

This following quotation is from an article that represents some of the strong views appearing on Chinese dissident websites: 'Can you establish your own political party? Can you set up a free labor union? Can you freely publish your political ideas? Organize groups? Demonstrate? As long as you don't have those few things, you don't have anything' (Backgo undated).

Other technologies used by dissidents are bulletin board sites and chat rooms. The number of postings to the forums has tended to peak around the time of important events. This is how the authors describe it:

> For example our observations indicate that traffic on these sites increased around the time of the diplomatic standoff that ensued after the April 2001 collision of a US reconnaissance aircraft and a Chinese fighter aircraft. In addition, in 2000, the number of postings on such sites surged around the times of the Taiwanese presidential election, Chen Shui-bian's inauguration, the permanent normal trade relations (PNTR) vote in the US congress, and the anniversary of the June 4, 1989, Tiananmen Square crackdown.
>
> (Chase and Mulveron 2002: 21)

Several dissident groups maintain their own BBS. The CDP established more than a dozen Chinese language BBS in May 2000. The Tibetan exile community also makes extensive use of BBS and chat rooms (Chase and Mulveron 2002: 22).

In some cases, postings have led to arrests. One example is Fu Lijun, 37, an assistant professor at Xinxiang Medical College in Henan, who was arrested in October 1999 for posting an article in a chat room detailing how Falungong could cure illness (Chase and Mulveron 2002: 22). Several other Chinese cyberdissidents are undergoing protracted legal battles over their online writings. A court has rejected appeals on behalf of four other people who were each given multi-year prison sentences for expressing themselves in cyberspace. Another online protestor, He Depu, was sentenced to eight years in prison for his activities in connection with the banned Chinese Democratic Party, and for posting supposedly subversive writings on the internet.

The harsh treatment of online dissidents has prompted protests from human rights groups. Robert Menard, the secretary-general of Reporters Sans Frontiers, expressed 'regret that the Chinese authorities have turned a deaf ear to the growing number of voices speaking out in China and abroad against their policy of cracking down on cyberdissidents' (Information Society News 18 November 2003). In November 2003, about 500 intellectuals in China and overseas signed a petition urging Beijing to free internet essayist Du Daobin, who was detained for allegedly 'subverting the state administration' (Lam 9 November 2003). According to Amnesty International's records, by January 2004, 54 people had been detained or imprisoned for expressing their opinions and downloading information from the internet, a 60 per cent increase on the November 2002 figures (Amnesty International 28 January 2004).

Chase and Mulveron offer us plenty of information on how Chinese students have used the internet extensively in the past five years: during the 1996 Diaoyu Islands dispute; in the aftermath of the accidental bombing of the Chinese Embassy in Belgrade in May 1999; following the murder in late May 2000 of Qiu Qingfeng, a Beijing university student; after the April 2001 collision of a US EP-3 surveillance plane and a Chinese F-8 fighter; and following the September 11 terrorist attacks on the United States (2002: 23). During the summer of 1996, renewed friction related to the longstanding dispute between China and Japan over the Diaoyu Islands prompted an outpouring of nationalistic sentiment and unauthorized public protests in China. The students used internet bulletin boards, chat rooms and email to organize protests and to disseminate information not carried in official media. Chinese leaders, worried that nationalistic outbursts could harm Sino-Japanese relations and even be directed against the regime for failing to be more assertive, temporarily shut down internet bulletin board sites at several universities in Beijing, and 'advised' the organizer of the protest campaign to leave the capital for 'vacation' in the remote Gansu Province. As for officials in Beijing, BBS offer potential political advantages:

> First, there is some evidence to indicate that government officials use popular sites such as the Strong Country forum to gauge opinion on a broad range of domestic and foreign-policy issues. Although the ultimate effects of the Internet remain to be seen there already been significant developments. Indeed in a forthcoming study of the Internet and the development of civil society In China, Guobin Yang argues that through the use of means such as chat rooms and bulletin board sites, Chinese Internet users 'are engaged in the discursive construction of an online public sphere'.
>
> (Chase and Mulveron 2002: 27)

Amnesty International has reported similarly:

> Nevertheless, Internet Activism appears to be growing in China as fast as the controls are tightened. Over the last year, there have been signs of Internet users acting increasingly in solidarity with one another, in particular by

expressing support for each other online. Such expressions of solidarity have proved dangerous as a growing number of people have been detained on the basis of such postings.

(Amnesty International Press Release 28 January 2004)

Furthermore, dissidents have a new method of communicating their views: email spamming. While spamming is a nuisance to most users, it can enable groups to transmit uncensored information to an unprecedented number of people within China, and to provide recipients with 'plausible deniability'. The publishers of two Chinese-language electronic magazines, Tunnel (Suidao) and VIP Reference (Da Cankao), have mounted the best-publicized and most sophisticated efforts. Tunnel, the first Chinese e-magazine, is published weekly and is reportedly compiled and edited largely within China, then sent to Silicon Valley, and finally mass-emailed back to the PRC from anonymous, US-based email accounts, such as nobody@usa.net (Chase and Mulveron 2002: 29). The editors claim that computers and the internet can be used to 'disintegrate the two pillars of an autocratic society: monopoly and suppression' (Chase and Mulveron 2002: 30). The VIP Reference magazine contains articles from Hong Kong, Taiwan and Western news sources that are not available to the public in China. The magazine declines to reveal how many mainland subscribers it has, but is reportedly sent to between 250,000 and 300,000 Chinese email addresses. The editors frequently change website addresses and use different email addresses every day to prevent Chinese security services from blocking distribution of their electronic publications. 'We are computer experts, and above all we like the concept of free speech. We are destined to destroy the Chinese system of censorship over the Internet. We believe that the Chinese people, like any other people in the world, deserve the rights of knowledge and free expression', said the Chinese editor of VIP reference (Farley 5 July 2004).

Moreover, users in China have translated Freenet, which allows people to exchange files over the internet through a shared network, like Kazaa and Gnutella, to Mandarin Chinese and have adapted it for distribution on a single floppy-disk. The reasoning behind this is to use it to share documents the government has been trying to censor, relating to the Tiananmen Square massacre and Falungong (Jardin 29 October 2002). File-sharing networks, using the same technology that gives American music and movie companies fits, can help dissidents communicate. Since networks like Gnutella and Kazaa have no central source, they are harder to turn off than centralized websites, chat rooms and BBS (Associated Press 27 August 2002). Also, numerous efforts are under way in the West to help Chinese web users get around China's censorship of the internet. In 2001, Bill Xia, who left China for the US in the 1990s, and some other US-based volunteers started Dynamic Internet technology. This allows a user inside China to access the internet, not through a system controlled by the government, but through a proxy server. Another idea is to present the Chinese authorities with so many proxy addresses that they would never be able to block them all. The US-based group, peacefire.org, has adopted this approach (BBC News 3 November 2004).

Often, the use of proxy servers becomes a controversy in itself. For example, Symantec has labelled a program that enables Chinese surfers to view blocked websites as a Trojan horse. Users of Norton Anti-Virus cannot access Freegate, a popular program which circumvents government blocks. Freegate lets users view sites banned by the Chinese government by taking advantage of a range of proxy servers assigned to changeable internet addresses (Indymedia 14 September 2004).

Nevertheless, dissidents, Falungong adherents and Tibetan exiles utilize websites for communication and motivation. The overseas branch of the CDP, Frank Liu, Falungong and the Tibetan exile community maintain particularly interesting and informative websites. The CDP web page includes links to organizational information, important CDP documents, a publicity department, an invitation to join the CDP and a variety of BBS forums. It includes a list of members of the CDP's national committee, with links to biographies of the most prominent members (Chase and Mulveron 2002: 32). Frank Siqing Liu, director and lone employee of the Hong Kong Information Center for Human Rights and Democracy, maintains a web page that features daily bulletins on arrests of dissidents and practitioners of Falungong, information on workers demonstrations and link updates from a variety of international sources such as the BBC and Radio Free Asia. Liu's website was hacked and the site was replaced with a message that said, 'this site sold to www.islam.org' (Chase and Mulveron 2002: 33).

The Falungong site (www.falundafa.org), which is bilingual, frequently updated, and well organized, contains messages from the cult's founder Li Hongzhi, links to 26 Falungong sites, a calendar of conferences and events, and audio downloads that enable practitioners to listen to Master Li's lectures from anywhere in the world. Also, Falungong has used text messages to send out thousands of messages, totally undetected. Meanwhile, the Tibetan community-in-exile and its supporting NGOs maintain a sophisticated and informative set of websites around the world. The main advocacy sites can be divided between those of officials, supporters and radicals (Chase and Mulveron 2002: 33–35).

Chinese dissidents' efforts towards political participation mirror both the unprecedented opportunities offered by the internet and communication technologies in general, and the actual pressing need to use them as an alternative information source and coordination network, providing the information citizens need to participate in political life and fight for a more open and democratic system.

According to Chase and Mulveron there are five key future trends in dissident use of the internet:

> First, in the short term, the Internet will require some human rights NGOs and advocacy groups to change their traditional focus on reporting arrests. Second, it will permit small groups and individuals with limited resources

to exert much greater influence than would otherwise be possible. Third, it appears likely that overseas dissidents, and perhaps even mainland dissidents, will engage in more email spamming campaigns in the near future. Fourth, dissidents may increasingly turn to emerging 'peer-to-peer' technology to exchange information. Finally, dissidents and other unauthorised organizations will try to find new ways to exploit the Internet's motivational and organizational potential.

(2002: 40)

In contrast to Chase and Mulveron's findings, administrative assistant Wang Yi, like many of her friends, is aware of the political nuances that can 'scupper' her surfing, but isn't too concerned about the situation:

The truth is most young people are just not that interested (in politics). I want to know how to improve my chances of a good job, others want to chat and find a boyfriend, and many just want to escape from hard studies by playing games like Counter-Strike. Personally, I know no one interested in cyber dissidents. It's just not part of our lives.

(Mackenzie 26 June 2003)

This view is also supported by Bobson Wong, an independent researcher based in New York, who says that the majority of Chinese web users are not looking to be cyberdissidents: 'These are not people itching to read CNN. There have been surveys done of internet users in China, and surveys reveal that most internet users in China trust the government' (BBC News 3 November 2004).

Despite the political use of the internet by dissidents, the Chinese government has been able to deal with it effectively using both low-tech and high-tech solutions. Installing advanced telecommunications infrastructure to facilitate reform has complicated internal security for the Chinese government. As Nina Hachigian puts it, the challenge for the regime is to 'prevent this commercial goldmine from becoming political quicksand' (Chase and Mulveron 2002: 46). President Jiang Zemin specifically threatened computer programmers, along with artists and writers, with stiff jail terms if they 'endanger state security' (Farley 5 July 2004).

The low-tech solutions employed by the Chinese authorities include the use of informers and surveillance, arrests of internet dissidents, the promulgation of regulations and, in some cases, the physical shutdown of network resources (Chase and Mulveron 2002: 49). The government also has the ability to search for key words and block sensitive email messages. Several hundreds of thousands of web pages, such as those devoted to Taiwan, the Falungong and foreign news, are blocked by the government. Chinese censors employ filtering technology to block and intercept emails to and from the country's nearly 80 million internet users. This is how Paul Mooney of the *Herald Tribune* describes it:

Beijing has become skilled at hunting down proxy servers that allow users to maneuver around firewalls ... Nor are Internet cafés safe havens any

longer for exploring the Internet. Cafés in some provinces are experiment-
ing with swipe cards linked to customers' national ID cards. One café
manager showed me a back room where a police-linked computer, connected to four spy cameras, monitored users.

(Mooney 23 April 2004)

Reporters Without Borders also reported that messages critical of the Chinese
government either never appear or are purged from popular chat rooms, while
Chinese law enforcement agencies track down and even jail the authors of the
messages. The most filtered topics are human rights, Taiwanese independence,
pornography, oral sex, SARS, the BBC and the Falungong movement.

A list obtained by the China Internet Project in Berkeley found that over
1,000 words, including 'dictatorship', 'truth' and 'riot police' are automatically
banned in China's online forums (*New Scientist* 24 November 2004).

The strategy of the security apparatus is to create a climate that promotes
self-censorship and self-deterrence. This is exemplified by the comments of a
public security official: 'People are used to being wary, and the general sense
that you are under surveillance acts as a disincentive. The key to controlling the
net in China is in managing people, and this is a process that begins the moment
you purchase a modem' (Chase and Mulveron 2002: xiii). No one knows exactly
how big China's internet police force is, although estimates run as high as
40,000 (Mooney 23 April 2004).

Low-tech solutions include shutting down internet cafés, an approach used
especially after the Lanjisu fire, but already present. For example, the Ministry
of Public Security authorities in Baoding City, Hebei Province, issued new regulations for 100 internet cafés in the city. The regulations announced a 'point
system', whereby a café 'allowing a customer to browse "reactionary" information will be deducted 10 points'. If a café loses 30 points within a single year, its
licence will be suspended for one year (Chase and Mulveron 2002: 60).

Following the fire at Lanjisu internet café in Beijing in June 2002, which
killed 25 people, the Public Security Ministry closed down 2,400 internet cafés
in that city, ostensibly for safety reasons. The government ordered all internet
cafés to augment their filtering software within weeks, and to keep records of all
users for a 90-day period. The software prevents access to 500,000 foreign websites, while those attempting to access these banned sites are automatically
reported to the Public Security Bureau. Internet police in cities such as Xi'an
and Chongqing can reportedly trace the activities of the users without their
knowledge and monitor their online activities by various technical means
(Amnesty International 26 November 2002). In March 2002, the authorities
introduced a voluntary pledge, entitled *A Public Pledge on Self-Discipline for
the China Internet Industry*, to reinforce existing regulations controlling the use
of the internet in China. Over 3000 Chinese internet business users have signed
the pledge, including the US-based search engine Yahoo! In protest against the
measures taken by authorities to control freedom of expression, information and
association on the internet, a group of 18 dissidents and intellectuals published a

Declaration of Citizens' Rights for the Internet on 29 July 2002 (Amnesty International 26 November 2002).

The net police closed almost half of the coutry's 200,000 internet cafés and installed surveillance software in the rest. In Liaoning province, where 40 per cent of the people who go online do so in internet cafés, software was installed in 7,000 cafés to track down web users' online movements and keep records of their names, addresses and ID numbers (*New Scientist* 24 November 2004).

As the Head of Human Rights in China Xiao Qiang explains:

> There are new subjects and topics being banned by the government and [internet] users need to come up with new key words or phrases for these topics. While the overall censorship is still effective, the Net has already created a bottom-up force and is constantly negotiating this new space with the old style, top-down censorship and propaganda regime. The transformative effect of the internet has already set China on an irreversible course towards greater openness and public participation in social and political life.
>
> (Greenberg 5 October 2005)

Government inspectors have checked up on 1.8 million cafés since the campaign began, seeking out those who let youths play violent games or access subversive foreign sites. In addition to the 1,600 cafés closed permanently, 18,000 have been shut down for 'rectification'. The government said that since the summer of 2004, 445 people have been arrested and 1,125 websites have been shut down (Sheriff 1 November 2004). Beijing does not actually label sites as 'blocked'. Instead, when a user clicks on a blocked site, the page will begin to download, slowly, and the user is redirected either to an error message or back to a Chinese search engine (McLaughlin 9 September 2004).

In one prominent example, email, mobile phone text messaging and world news coverage that made it to China in part via the internet are widely credited with forcing the central government to go public with the facts and scope of the 2002–3 SARS epidemic.

In addition to traditional methods of control, the Chinese authorities have also made use of high-tech countermeasures, such as blocking websites and email, government-sponsored hacking, monitoring and filtering of email and online propaganda, denial, deception and misinformation (McLaughlin 9 September 2004: 61).

Nevertheless, the most effective line of defence in China's internet security strategy is the use of bureaucratic regulations to shape the market environment and the incentives of key participants in ways favourable to the state's interests. Since 1995, the Chinese government has promulgated a blizzard of rules governing nearly every aspect of the internet market. In particular, the 1997 Public Security Bureau regulation entitled 'Computer Information Network and Internet Security, Protection and Management Regulations' places most of the onus for monitoring, reporting, and preventing anti-regime use of the internet on domestic providers (McLaughlin 9 September 2004: 57). As a result, ISPs have implemented certain self-censoring policies to avoid trouble with the authorities. According to a 24 January 2000 Reuters story, ISP employees monitor chat rooms and bulletin boards, ferreting out risky political commentary, foul language and unwanted advertisements. The most recent regulations on the use of the internet were promulgated by the Ministry of Information Industry in January 2002. These regulations require ISPs to maintain detailed records about their users, install software to record email messages sent and received by their users, and send copies of any emails that violate PRC law to the appropriate Chinese governments departments (McLaughlin 9 September 2004: 61).

Also, authorities at various times have blocked politically 'sensitive' websites, including those of dissident groups and major foreign news organizations such as the Voice of America, the *Washington Post*, the *New York Times* and the BBC. In September 2002, they even banned the search engine Google. Google is hugely popular among China's internet users because of its wide-ranging capacity. A search in English for former President Jiang Zemin's name turns up links to 156,000 websites mentioning him as opposed to a search on Sina.com, which turns up 1,600 mentions of Jiang. The Chinese-language service of Yahoo turns up just 24, largely because Yahoo has offices in Beijing and follows the censorship restrictions. Also, in the case of Google, even if a site is made unavailable by a government or an ISP, Google makes the site available through the cache of pages it photographs as it crawls the web. The fact that numerous news sites especially are blocked by the Chinese authorities indicates that they are not prepared to let access to information – a basic ingredient for citizen participation and democracy – be a democratizing force in the People's Republic of China. The dissident groups fighting against censorship and for democracy in China are prosecuted by the state, spied upon, arrested and then given prison sentences of up to 15 years.

Matters became worse for Chinese surfers just before the November 2002 Communist Party Congress. State media quoted Jiang in August as telling propaganda officials to create a 'sound atmosphere' for the meeting. Michael Robinson, chief technical officer of Beijing-based Clarity Data Systems, was quoted as saying at the time: 'this is a serious escalation. They are not acting as administrators. They are acting as hackers. They are impersonating authority they don't actually have' (Bodeen 3 September 2002). The move appeared to have been ordered by public security authorities and implemented locally via internet servers run by the country's fixed phone giant China Telecom. Nevertheless, the battle for Google between Chinese authorities and the country's millions of internet users ended in defeat for the web censors. The censors allowed access to be restored after a complete block on the site for almost ten days. However, they also put on filters that would refuse searches for President Jiang and other politically sensitive queries (Gittings 17 September 2002).

For internet service providers, blocking certain sites in China is done as easily as flipping a light switch. As much as 80 per cent of the country's internet traffic flows through ChinaNet, a subsidiary of China Telecom. 'On balance, we believe that having a service which links that work and omits a fractional number is better than having a service that is not available at all. It was a diffi-cult trade-off for us to make, but the one we felt ultimately serves the best inter-ests of our users located in China', Google said in its blog. The organization Reporters Without Borders has accused Google of pandering to Chinese inter-ests and filtering its Chinese-language site (Singer 30 November 2004). Bill Xia, of Dynamic Internet Technology, a research firm striving to defeat online cen-sorship, suspects that Google is cooperating with the Chinese government's cen-sorship efforts to smooth the way for expansion plans that could help them boost future profits (Associated Press 25 September 2004). Although China no longer blocks Google in its entirety, a Chinese user of Google can potentially have a very different experience than one from another country, due to China'a content filtering practices. Chinese Internet users' access to Google is filtered for spe-cific keywords, and this filtering disrupts both Google searches as well as access to the Google cache. The isolation and filtering of the text string 'search?q=cache' suggest that the Chinese state knows how to access Google's cache, which can be used as an ad hoc form of censorship circumvention. The state has taken steps to limit the cache's power accordingly. The fact that other search engines' cache functions are not filtered suggests that China has deliberately targeted Google for filtering (Open Net Initiative 30 August 2004).

Google, the 'do no evil' company, has been condemned for its decision in 2006 to comply with the Chinese government's wishes to censor access for Chinese users. Remarkably, on 2 February 2006, as investors were stunned by a quarter earnings report falling far short of expectations, $13 billion was wiped off its stock value (12 per cent). The company is also pressured by many tradi-tional industries:

> telecommunications companies do not like its plan for free internet phone calls, book publishers and newspapers have filed a lawsuit to try to prevent it from digitising library materials, governments are worried about its satel-lite imaging service Google Earth and provacy advocates have a growing list of concerns about everything from its email service to its destop search function, both of which make it easier for hackers or government agencies to gather information about individuals without their concern.
>
> (Gumbel 2 February 2006)

To come back to Chinese cyberdissidents, most of the government's attempts to prevent the viewing of banned websites use multiple layers of filtering, ranging from the ISP to the network carrier to the non-technical aspects of web surfing in China (e.g. registration with the police, observation by internet café workers). In addition, blocks on websites are often temporarily removed when high-level foreign delegations visit China. During the October 2001 Asia Pacific Economic Cooperation (APEC) meeting in Shanghai, for instance, Beijing per-mitted access to several websites that are normally blocked, including those of the Washington Post and CNN. This was presumably intended to avoid embarrassing international media reports which would burnish China's image, in order to make a good impression on visiting world leaders. As soon as President Bush and other world leaders left Shanghai, however, the websites were once again blocked (Chase and Mulveron 2002: 64).

However, despite such efforts to block sensitive sites, the China News Digest (CND) is at the forefront of the promotion of one of the most potent weapons used in fighting website blocking: proxy servers. A proxy server sits between a client application, such as a web browser, and a real server. In addition to evading blocking, proxy servers are also used to improve performance and filter requests. The CND website contains a guide to using proxy servers to circumvent firewalls and internet censorship, as well as a call for volunteers to provide CND with additional proxy services. However, Chinese authorities have redoubled their efforts to block access to popular proxy servers (Chase and Mulveron 2002: 66). For the moment at least, Beijing has the upper hand in the website-blocking battle. Chase and Mulveron cite one activist who recently visited the mainland: 'the authorities have become much better at finding and blocking proxies in China; I was there for eight days and experienced eight days of Internet black out' (Chase and Mulveron 2002: 69, interview with a Chinese activist).

Zittrain and Elderman, who did an empirical analysis of internet filtering in China, tracked 19,032 websites that were inaccessible from China on multiple occasions, but remained accessible from the United States. The authors concluded: (1) that the Chinese government maintains an active interest in preventing users from viewing certain web content; (2) that it has managed to configure overlapping nationwide systems to effectively – if at times irregularly – block such content from users who do not regularly seek to circumvent such blocking; and (3) that such blocking systems are becoming more refined even as they probably require more labour and technology to maintain than did cruder predecessors (23 March 2003).

Furthermore, Chase and Mulveron argue that there is some evidence to suggest that the Chinese government, or elements within it, have engaged in hacking of dissident and anti-regime computer systems outside of China. Because of the difficulty in establishing official culpability for hacking attacks without additional evidence, governments can claim a reasonable measure of plausible denial in these cases. This is how the authors describe it:

> The Chinese origin hacking attacks that occurred against Taiwan in August 1999 and against Japan in February 2000 are examples of incidents in which government culpability, either limited or complete, is difficult to determine solely on the basis of the intrusion data. Stronger evidence exists to support the conclusion that the Chinese government or elements within it were responsible for one or more of the China-origin network attacks against computer systems in the United States, Australia, Canada, and the United Kingdom.
>
> (Chase and Mulveron 2002: 71)

In October 2003, it was revealed that a Beijing-based internet company was implicated in creating a program specifically designed to spy on computers of pro-Tibet dissidents. The attempt to spy has been done through sending innocuous-looking messages, purportedly from officials of the Tibetan Government-in-Exile (TGIE) and members of Tibet Support Groups (TSG), with subject matters listing current developments, including a recently held Fourth International Tibet Support Group Conference in Prague. Once attachments to such emails were opened, they planted a Trojan horse on the computer, which made its content accessible to the Internet Group in China. These emails came from a Chinese IP and contained attachments of a Trojan horse that could potentially do anything.

This was not the first time that organizations in China have tried to penetrate into the network system of TGIE. Jigme Tsering of the Dharamsala-based Tibetan Computer Resource Centre (TCRC) has revealed that there have been repeated attempts in the past to infect TGIE computers with viruses in order to obtain information. In an interview to the UK internet news site *The Register* in September 2002, Tsering warned that Tibet supporters were being targeted by an unnamed virus which was designed to fool the unwary by posing as an email from the Dalai Lama's office (Save Tibet 28 October 2003). It is no secret that, in addition to forcing ISPs and internet cafés to use filtering software, government minders hand-delete individual messages from discussion boards and change the domain-server records of forbidden sites, so that visitors are routed to authorized pages.

Another incident indicates how the Chinese government deals with dissidents' use of the internet. Human Rights Watch reported that in May 1998 the Ministry of Labour and Social Security installed monitoring devices at the facilities of ISPs that could track individual email accounts (Chase and Mulveron 2002: 82). In January 2000, Liu Ming, the younger sister of student leader Liu Gang, wrote an indictment against the Changchun City Public Bureau, which was then disseminated abroad via the internet by Leng Wanbao, a noted dissident in Changchun City, Jilin province. Leng was picked up and interrogated for three hours by the police, who knew about the activity immediately (Chase and Mulveron 2002: 82). Chinese authorities have also detained Du Daobin, a civil servant, whose essays are banned by Beijing on the internet, on charges of subversion. Du who works for the municipal medical reform office, signed an online petition calling for the release of fellow cyberdissident Liu Di, a female psychology student from Beijing Normal University who was detained in the capital in November 2002 (Reuters 31 October 2003).

In November 2003, a court turned down the appeals of four internet dissidents who were sentenced to up to ten years in jail for posting their political views. They were arrested in March 2001 after they set up the 'New Youth Association', an academic study group that discussed China's growing social problems (*South China Morning Post* 11 November 2003). Similarly, Li Zhi, a 32-year-old local official from Dazhou, has been charged with conspiracy to subvert state power, according to the Human Rights in China (HRIC) group. Li frequently expressed his views on online bulletin boards and in chatrooms, and he has also been accused of communicating with overseas dissidents. Liu Qing, the president of HRIC has commented that '[m]onitoring email and internet chat rooms is an unacceptable invasion of privacy, and a reprehensible method of gathering evidence for prosecution of a political crime' (BBC News 24 September 2003).

The Chinese forums

> use a system of filters that enable them to sort the messages to two categories: those containing banned words and the rest. Messages in the first category are systematically blocked ... Site Web Masters are supposed to check these blocked messages to establish whether they really need to be censored. But in fact, it is very rare for a message that has been filtered out to be manually restored to the forum.
>
> (Ching 22 May 2003)

Some analysts argue that, at least in the short to medium term, the spread of the internet will tend to benefit authoritarian regimes at the expense of dissidents and pro-democracy activists (Chase and Mulveron 2002: 87). As Kalathil and Boas observe, for example, China and other authoritarian states have responded effectively to the dissident challenge by implementing a combination of reactive measures, including blocking websites and jailing activists, and proactive policies, such as distributing propaganda online and proffering e-government services (Chase and Mulveron 2002: 87). However, China's proactive efforts to use the internet to bolster regime power have thus far produced only limited results. For example, although Beijing has actively promoted a 'government online' plan, a recent survey of e-government initiatives around the world found that China ranked 83rd out of 196 countries (Chase and Mulveron 2002: 88).

While Beijing has done a remarkable job of finding effective counterstrategies to potential negative effects of the information revolution, the scale of China's information-technology modernization would suggest that time is eventually on the side of the regime's opponents. Nina Hachigian predicts that 'control over information will slowly shift from the state to networked citizens' leading to potentially 'seismic' changes (Chase and Mulveron 2002: 89).

Interestingly, this is how Chase and Mulveron put it:

> The government will not lose the upper hand soon though, because the government's strategy is also aided by the current economic environment in China, which encourages the commercialization of the Internet, not its politicization. Thus the Internet, despite the rhetoric of its most enthusiastic supporters, will probably not bring 'revolutionary' political change to China, but instead will be a key pillar of China's slower, evolutionary path toward increased pluralization and possibly even nascent democratization.
>
> (2002: 90)

The fact that the internet is so hard to control has also given it a role in making other media more open. According to Liu Qing, a prominent dissident who left China after his release from jail in 1992, and who is now chairman of the US-based Human-Rights in China, 'no matter how sophisticated its technology there is no way the government can fully control the internet. That's why in China these days you can see all kinds of organizations and activities springing up, moving the country towards real change' (Luard 30 January 2004). The latest news of China's censoring techniques is that China is expanding its censorship controls to cover text messages sent using mobile phones (Lim 2 July 2004).

By the end of October 2004, China had more than 45 large blog-hosting services. Any tech-savvy user can download and install blogging software themselves, bypassing the controls. Blogs play an important role in republishing information as quickly as it is banned from official websites. For example, China's most influential bulletin board Yitahutu (the site had more than 300,000 registered users and 700 discussion forums) was closed down by the net police. After the closure, all the major university bulletins were instructed to delete any discussion of the event. Even the name was censored from Chinese search engines. But the net police found it harder to purge discussion, because bloggers found euphemisms for the event despite key word filtering (*New Scientist* 24 November 2004).

Bloggers in China have had email messages telling them to register or face sanctions. And, according to Reporters Without Borders, one blogger who contacted the Shanghai police to register was told there was no point in registering, as independent blogs would not be granted permission to continue (*The Enquirer* 7 June 2005). By January 2003, China had about 2000 bloggers when, without warning, the Chinese government blocked all access to blogspot.com, the server that hosts all blogs registered on blogger.com. The net police do not make the reasons for such actions public, but Chinese bloggers point out that DynaWeb, an anti-censorship service run by overseas Chinese, had been using a blog on blogspot.com to publish proxy addresses that allowed users to get around the Great Firewall. The authorities' blanket blockade affected all China's bloggers, leaving them suddenly unable to reach their journals (*New Scientist* 24 November 2004).

A study from the Harvard Law School found that the Chinese government has become increasingly sophisticated at controlling the internet, taking a multi-layered approach that contributes to precision in blocking political dissent by blocking just specific references to Tibetan independence without blocking all references to Tibet. John Palfrey, one of the researchers of the study, sees China as the most successful country in the world to manage to filter the internet despite the fast changes in technology. The testing determined that, though some dissidents complain that email newsletters sent in bulk are sometimes blocked, individual messages tend not to get filtered. This is because individual internet service providers, cybercafé owners and discussion forum moderators deploy additional blocking under threat of penalties by the state apparatus (Associated Press 14 April 2004).

The Chinese case is closely related to this work's five research parameters – democracy, participation, power, globalization and social movements – because these movements are globalized, involve groups that seek participation and power (Falungong) or, in some cases, participation, democracy and power (CDP and the Tibetan exile community). The information restricted in the Chinese state offline is available in cyberspace, subverting the national boundaries that have, in the past, helped the Chinese government to control access to information. Despite the fact that the Chinese authorities have been competent in internet filtering and arresting cyberdissidents, total control of information spreading through the internet ultimately proves impossible. This is why cyberdissidents will, in the future, have more opportunity to voice their viewpoints and coordinate their activities through the net, demanding more participation, more power and definitely more democracy from the Chinese state.

Internet censorship internationally

In its 2004 report 'Internet Under Surveillance', the RSF gave positive grades only to Japan and Taiwan for internet freedom. The US was considering allocating $100 million to thwart internet censorship by authoritarian regimes with the introduction of a bill that would establish an Office of Global Internet Freedom, to foster development of censorship-busting technology for users in such countries as China and Saudi Arabia. The bill would allocate $50 million each for 2003 and 2004. The bill says: 'With nearly 10 per cent of the world's population now online and more gaining access each day, the internet stands to become the most powerful engine for democratization and the free exchange of ideas ever invented' (Wagner 3 October 2002). The bill cites a catalogue of censorship techniques: surveillance of email and message boards, blocking content based on keywords, blocking individuals from visiting proscribed websites, blacklisting users and wholesale denial of internet access. An attempt to overcome this is with Six/Four, developed by the Hacktivismo Group, was planned as a full-scale peer-to-peer platform for all internet activities.

Ironically, in June 2002, with the restructuring of the US government, Attorney General John Ashcroft scrapped the guidelines that govern the FBI's conduct, allowing the bureau to monitor websites, public gatherings and religious institutions that are not under criminal investigation. Ashcroft also said that the government would photograph and fingerprint up to 100,000 foreigners entering the country from Arab and Muslim countries. Marc Rotenberg, director of the Electronic Privacy Information Center, commented: 'I think we've reached the point in the debate where we need to ask larger questions about where this administration is taking the US government. Someone needs to apply the breaks or the United States will become a "police state"' (Manjoo 7 June 2002). The Congress approved a bill in November 2003 that expands the reach of the Patriot act, shifting the balance of power away from the legislature and the courts, while expanding the power of the FBI to subpoena business documents and transactions without first seeking approval from a judge. Under the

Patriot Act, the FBI can acquire bank records and internet or phone logs simply by issuing itself a so-called national security letter, saying the records are relevant to an investigation into terrorism. The target institution is issued an order and kept from revealing the subpoena's existence to anyone, including the subject of the investigation (Singel 24 November 2003). A new law officially called the Domestic Security Enhancement Act of 2003, called PATRIOT II by privacy advocates, calls for the creation of a terrorist DNA database, eases laws pertaining to search, seizure and admissible evidence, and would allow the attorney general to revoke the citizenship of any resident who provides 'material support' to terrorist groups (Delio 3 April 2003). Similar concerns are raised in an EU report released by the European Commission, which argues that after the September 11 terrorist attacks in the US, 'many governments enhanced their surveillance powers, but at the risk of affecting privacy'. Phillipe Busquin, the European Union's commissioner for research, has stated that 'citizens are not prepared to let privacy be one of the casualties in the war on terrorism' (Wenzel 18 November 2003).

An example of the use of state powers against privacy arose in September 2002, when the University of California had to reconsider their decision that would have forced a student activist group to remove from its website a link to a guerilla group accused of being a terrorist organization. The campus activist group, known as Burn, provided a link which directed visitors to the official site of the Revolutionary Armed Forces of Colombia (FARC), one of the 34 groups on the US government's list of foreign terrorist organizations. Gary Ratcliff, UCSD University Centers director, sent the initial cease-and-desist letter to the Che Café Collective, the university group that sponsors Burn, which cites a section of the USA Patriot Act that deems it unlawful for any US citizen to provide 'material support or resources' to foreign terrorist organizations. Civil rights groups have heavily criticized the Patriot Act, claiming it violates constitutional rights (Wenzel 18 November 2003). A UCSD student and member of the Collective, who says her name is Allie Katz, noted that Burn would not take down the link if the university decided to stand by its initial demand. 'We see this as a free-speech issue', she said. 'Merely having a link doesn't constitute material support.' Ratcliff responded that 'the academic computing department has found that some outside groups have Unix accounts on the Burn server. We're not sure if FARC does, but that's why we'd like to talk further with the students' (Asaravala 28 September 2002). In April, UCSD officials had succeeded in having the Groundwork Collective, Burn's previous sponsor, remove a link to the PKK. If the university stands by its initial demand and succeeds in having the FARC link removed by way of the PATRIOT Act, the decision could set a precedent for other public institutions looking to eradicate controversial links and sites. Another story relating to academia in the US is the website www.campus-watch.org, set up by Middle East Forum, a think-tank set up 'to monitor the attitudes of American professors and universities toward Islamic fundamentalism'. What gave the site instant notoriety was its posting of extensive dossiers on eight dubious professors. The students were invited to 'inform' on any other of their teachers who should join the 'treacherous octet' (Sutherland 7 October 2002).

Moreover, in contrast with China, which targets political freedom, Saudi Arabia clamps down on personal freedom, showing greater concern for personal morality than political subversion. According to a Harvard Law School study of the Saudi Arabian case, the Saudi government is keeping its subjects from viewing sites about drugs, women and rock 'n' roll. The Saudi case is more typical than China's, although a few regimes like China exist – notably Vietnam and the United Arab Emirates, which actually attempt to filter their national internet traffic wholesale (Shachtman 18 July 2002). Ben Edelman, one of the analysts behind the report on Saudi Arabia, has commented: '[i]f you are trying to get into the club of rich, industrialized countries, filtering is not something you want to be seen as doing. But you need to be wealthy enough to afford a filtering system. So it's not happening in Africa, for example. There's no money for it' (Manjoo 7 June 2002). For the study, Jonathan Zittrain and Ben Edelman tested 64,000 websites, with the full collaboration of the Saudi government. Most other countries are not willing when asked. The Saudis are open about their censorship of the web. If a site is blacklisted, the user is directed to a web page that informs him or her that access is denied, in contrast with China, where a surfer simply gets an error message (BBC News 31 July 2002).

Radical Islamic websites are not the only contribution of Islamic faith on the web, as more moderate groups struggle to break the monopoly of the state in setting the political and social agenda. These sites aim particularly to attract younger Muslims by offering a modern interpretation of Islam, disseminated by modern or perhaps postmodern technology. Examples of sites include the Cairo-based Islamonline website (www.islamonline.net), set up by Egyptian Islamist intellectuals with a 24 hour news service providing a wealth of information with the objective 'to work for the good of humanity and to support principles of freedom, justice, democracy and human rights' (Abdel-Latif December 2004). Also, the Al Shaab newspaper, the mouthpiece of the Islamist-oriented Labour Party, went online after being banned, with the editor publishing fierce criticism of the regime. This resulted in censorship from the state and the site was hacked many times. Another newspaper of the Muslim Brotherhood, shut down by the Egyptian government, went online (www.ikhwanonline.com) connecting to audiences home and abroad, and fielding 54 candidates as independents (Abdel-Latif December 2004).

Text messaging has also been a more recent element of Arab dissent during the past 15 years. Saudi exiles and Islamic activists waged an underground war of faxed pamphlets during the early and mid-1990s. Satellite television channels transformed the images and ideas available to Arab viewers during the same period. More recently, CDs, DVDs and the world wide web have dominated underground political publishing in the Gulf.

As Coll writes:

> The technology also helps democratic organizers who are often overmatched by the Gulf's authoritarian governments. In a region where formal

political parties are banned but loose political societies are often tolerated, text messaging allows organizers to build unofficial membership lists, spread news about detained activists, encourage voter turnout, schedule meetings and rallies, and develop new issue campaigns – all while avoiding government-censored newspapers, television station and websites.

(29 March 2005)

He mentions the example of Kuwaiti women organizing protests for voting rights, and said they had been more effective during their 2005 campaign than during their last serious effort five years ago, because text messaging had allowed them to call younger protesters out of schools and into the streets (29 March 2005).

And things can get really dangerous for cyberdissidents in these countries. Massoud Hamid, a 29-year-old Syrian journalism student was sentenced to three years in prison for posting photographs on a website. The photos showed a peaceful Kurdish demonstration outside UNICEF's Damascus headquarters. Hamid has already spent 14 months in prison (Kiss 14 October 2004).

In a world-wide trend of repressive governments cracking down on internet journalists and dissidents, Tunisia's crackdown on cyber-dissidents took an ominous turn, with the arrest and detention of journalist Zouhair Yahyaoui, founder and editor of the online news site TUNeZINE. Better known under the pseudonym Ettounsi ('The Tunisian'), Yahyaoui was charged under Clause 2 of Article 306b of the Tunisian criminal code for 'knowingly putting out false news' and also for 'stealing' internet connection time at a local cyber café where he was working. Yahyaoui set up the site in July 2001 to put out news about the fight for democracy and freedom in Tunisia (McGrath 13 July 2002). He published opposition material online and was one of the first people to circulate a letter from his uncle, Judge Mokhtar Yahyaoui, to President Ben Ali, criticizing the country's legal system. Between May 26 and 28 2002, TUNeZINE (www.tunezine.com) organized an online forum on a recent government referendum and the state of the opposition, which drew a large number of participants. The TUNeZINE website, which is hosted in France, had been censored by the Tunisian authorities from the outset, but each week a list of 'proxy' addresses has been available so Tunisians could get around the blockage and access the site. Reporters Without Borders denounced as 'scandalous' Yahyaoui's jailing for two years and four months. His lawyer, who visited him in prison, said he had been slapped and hit on the head while being interrogated. He was undressed and tortured three times by being made to hang by his arms with his feet barely touching the ground. After the last session of this, he revealed the password to his site, which enabled the authorities to block public access to it (McGrath 13 July 2002).

Iran also leads in crackdowns on internet journalists and dissidents, according to a report from the *Guardian* (Europemedia 10 March 2003). The report specifically cites the arrest of journalist Sole Sa'di following an online article he had written, criticizing supreme leader Ayatollah Ali Khamenei and citing the arrest of Mohamed Mohsen Sazegra, the manager of news site alliran.net, as a sign that the clerics who run the society will not put up with the relative freedoms that flourish on the internet. Moreover, a number of internet service providers have been shut down, leading to the resignation of the head of the national association of Iranian ISPs in protest (Europemedia 10 March 2003).

Blogging has served as an outlet for dissent. According to Hossein Derakhstan, a 28-year-old blogger, '[u]ntil there is a free press in Iran again weblogs will flourish. In the last few years about 90 per cent (pro-democracy) newspapers in Iran have been shut down. So people have turned to the Internet to get news' (Delio 28 March 2003). But the relative tolerance of blogging may be coming to an end, since journalist Sina Motallebi was arrested in Tehran for blogging. According to the Islamic News Agency, he has been charged with 'undermining national security through cultural activities' through the content of his blog, as well as in his other writings and the interviews he gave to foreign media outlets (Delio 28 March 2003). His arrest brought fear to other bloggers, as people have stopped blogging or have censored their blogs by removing any posts that might offend.

Nevertheless, the socio-cultural importance of blogging is substantial. 'The blogs show that [Iranians] are carrying new values and promoting new lifestyles. Older generations try to hide their personal feelings and opinions from others. Individuality, self-expression and tolerance are new values which are quite obvious through a quick study of the content of Persian web logs', Derakhshan said (Delio 28 March 2003). Similarly REDCAT, a theatre in Los Angeles, has been screening a series of films from Iran's hidden cinema. Underground film-making means that as more filmmakers hit the streets with digital camcorders in their backpacks, the Iranian government may have a harder time tracking them down:

> With the advent of digital modes of recording, a lot of things we thought we knew about repressive regimes and modes of censorship of cinema are not true anymore. With a pocket digital camera you can record professional images, and you can have an editing system at home, so you don't need a permit. And you can walk outside of the country and present these images in the world.
>
> (Berenice Reynaud quoted in Silverman 6 February 2004)

Iran's hardline clerics publicly 'disqualified' more than 2,500 reform candidates for Parliament, but could not shut down either the 30 major political websites which are accessible in Iran or the 20,000 Iranian blog sites. The fundamentalists' landslide victory in Iran's summer 2004 elections might have disheartened Western observers, but these alternative channels of communications show signs that the internet has become the most successful way to work around oppression (Thomas, fwd from Smygo list).

Hundreds of Iranian online journals have been protesting against media censorship by renaming their websites after pro-reformist newspapers and websites that have been banned or shut down by authorities. It is thought that the number of Iranian blogs is now between 10,000 and 15,000. Some recent reports have suggested that Iranian authorities are considering the creation of a national intranet – an internet service just for Iran – in an attempt to separate from the world wide web (BBC News 22 September 2004).

Iran has a thriving community of around 46,000 bloggers, but the government has persisted in persecuting people connected with the pro-democracy writing (Kiss 18 March 2005). An imprisoned Iranian blogger, Mojtaba Saminejad, was arrested in November 2004 after writing about the arrests of three fellow bloggers. He was released on payment of £30,000, but was rearrested for relaunching his blog. In May he went on a hunger strike. While in prison his blog was hijacked by hackers connected to the Hezbollah group. The international group Committee to Protect Bloggers organized a day of protest for the blogger and offered campaign banners and advice on writing letters to the Iranian government.

In Iraq, The blog Iraqthemodel.blogspot.com and reportedly 60 others are bringing to light events and public opinion in Iraq that are not commonly aired by conventional media in the United States and elsewhere (Koprowski 6 October 2004). While American bloggers have challenged the veracity of major media reports, their Iraqi counterparts are creating a true, free press, online, in their homeland, for the first time in that country's modern history, using internet technology. The blogs are getting 3,000 to 6,000 visits per day – up to 200,000 visitors per month. Another blogger, Zeyad, of the HealingIraq.blogspot.com, reported on the demonstrations against terrorism last December, which went unreported by the international and local media. Sam, a blogger who runs Hammorabi, reported on his site that Dutch troops who are participating in the coalition have received death threats reaching the top levels of command. Some of these threats came as telephone calls to their private phones, as messages to their mobile phones, or to their emails.

Moreover, at least pre-war, the internet was too expensive for Iraqis, with academics and government officials allowed to open email accounts on a monthly fee of $25 dollars a month, in a nation where a university professor is lucky to earn $120 a month. Censorship was also present. For example, a search entry of 'Israel' received a response of 'Your access has been denied'. A reporter for *Wired* turned the tables when he hacked into Saddam Hussein's email account through his 'send mail to' link on the official Iraq website (ABC News).

In Afghanistan, where the internet is too expensive for the public to use, viewing is censored by running Net Nanny, a program that allows administrators to block sites, chats and newsgroups, as well as monitor online activity. Despite these measures, motivated users will always find ways around technical mechanisms intended to control their web browsing. Technologies aimed at controlling internet content tend to fail in two different ways. 'Not only are they unable to reliably block access to the forbidden fruit, but they also tend to inappropriately block innocent materials, often of social, health, political or other major importance' (Weinstein 16 December 2002).

In Pakistan, the military government, as a matter of policy, blocks websites. For instance, they have curtailed access to a Washington-based news website *The South Asia Tribune*, which was launched in 2002 by journalist Shaheen Sehbai. It was blocked by the Pakistan Internet Exchange (PIE), a subsidiary of the Pakistan Telecommunication Company which provides internet bandwidth in Pakistan. Since the beginning of 2003, the PIE has denied access to objectionable websites, and has also banned internet telephony and voice/chat websites. But technology surpasses such bans, as internet users in Pakistan are logging on *en masse* to proxy servers, thousands of which are available on the world wide web (Mustafa 4 June 2003).

North Korea, in keeping with the country's isolationist, agrarian agenda, has simply rendered internet access illegal. Computers are rare and internet access is almost non-existent, yet North Korea is suspected to have a military academy specializing in electronic warfare, turning out 100 cybersoldiers every year for nearly two decades (McWilliams 2 June 2003). Graduates of the elite program at Mirim College are skilled in everything from writing computer viruses to penetrating network defences and programming weapon guidance systems. In May 2003, South Korea's Defence Security Command raised the issue at a cybersecurity seminar. In fact, the North Korean government would be grossly negligent if it failed to beef up its information warfare capability, because South Korea, one of the most wired nations in the world, makes no secret that preparing for war is a top military priority (McWilliams 2 June 2003).

South Korea, the world leader in broadband access, with more than half the population online, is not above net censorship. In July, two students were arrested for posting material online that mocked political candidtates and others for promoting communism via the web. The government blocked a reported 18,000 web pages last year, according to Reporters Without Borders.

Another example of conflict between dissidents and the government occurred in Ukraine, when Ukraine's successor to the KGB, known as the SBU, wanted to take over the top-level domain name. Dmitry Kohmanyuk, a network systems administrator who officially has control of the domain name and runs it with a group of volunteers in the US, said at the time that he would give it up, but only if the new system was based on the internet principle of non-discrimination, and was open to everybody. The reason he did not trust the SBU was because in 2000 Georgiy Gongadze, the founder of a feisty political site Ukrainian Truth, was found murdered outside Kiev (Barton 22 June 2001).

Also, Burmese dissidents are operating more and more from neighbouring countries such as India, where they use the net to campaign against the military government. Aung Naing, editor of the Dhaka, Bangladesh-based Network Media Group commented that 'online independent news groups have become the window for the international community to peer inside this opaque country'. Burmese dissidents complain that their email systems are frequently attacked by viruses (USA Today).

In South Africa in July 2002, internet professionals accused the government of trying to hijack the world wide web and petitioned President Mbeki not to sign a new bill that would allow the government to take control of the registra- tion and administration of internet domains (website names, addresses and space), and give it free access to information stored on the web (Wagner 3 October 2002). Also, in Zimbabwe, the country's only independent newspaper was relaunched online, despite being banned by the government. Production stopped when armed police raided the newspaper's office in Harare. Staff were ordered from the premises and computer equipment was confiscated. Just days after arrest warrants were issued for 45 *Daily News* journalists, Ngunjiri Nderitu, web editor of the newspaper, hoped that the internet would allow them to reach their readership in Zimbabwe. *Daily News* staff moved to Johannesburg to avoid government action, and their website remained active throughout the difficulties (Dotjournalism 18 November 2003).

Another example is Cuba, where, despite the strict censorship of the net (the Cuban government initially banned the sale of computers to the public), the government launched a website in September 2002 to refute US charges that it sponsored terrorism, and to seek support in the United States for the release of five Cubans imprisoned there for spying. Cuba is still on Washington's blacklist, along with Iran, Iraq, Libya, North Korea, Syria and Sudan, for providing a haven for Basque separatist group ETA and supporting Colombian guerillas. The site www.antiterroristas.cu published an interview with Noam Chomsky, who criticized his country for using the term 'terrorism' only for acts of violence against the United States.

Vietnam presents another interesting case in point, where the government spies on internet café customers to prevent them from accessing documents it considers politically and morally objectionable. Vietnamese people living abroad, as well as dissidents inside the country, use the internet to circulate doc- uments critical of the government. There have been cases where people were arrested for posting on the web. Pham Hong Son, for instance, was arrested for translating and circulating on the net an article about democracy from a US State Department internet site. Internet service providers in Vietnam are responsible for filtering websites, but internet café owners are required to monitor customers as well. The prime minister also issued a directive prohibiting all citizens, except top Communist party and government officials, from watching international satellite TV (Wired.com 26 June 2002).

A June 2004 report titled *Internet Under Surveillance*, by Reporters Without Borders, summarized the censorship tactics of authoritarian regimes. These include blocking sites, targeted filtering, modified mirrors (authorities in Uzbek- istan change the content deemed unfavourable), prohibiting web-based emailing and owning ISPs, forcing cybercafé users to show IDs, and banning access and equipment (Scheeres 22 June 2004).

Conclusion

This chapter has examined empirical examples of sociopolitical cyberconflict and dissidents' use of the internet against authoritarian governments. In particular, it has looked at the anti-globalization movement and Chinese dissidents' use of the net, as well as the Chinese government's counterstrategies. The remainder of the chapter included examples of dissidents fighting online against other authoritarian regimes. This showed how widespread both state censorship activities and attempts to circumvent these controls have become.

The integrated theoretical framework, proposed in the section on pages 86–93 of Chapter 2, urges a focus on the following parameters.

Sociopolitical cyberconflicts

The impacts of ICTs on: a) mobilizing structures (network style of movements using the internet, participation, recruitment, tactics, goals); b) framing processes (issues, strategy, identity, the effect of the internet on these processes); c) political opportunity structure (the internet as a component of this structure); d) hacktivism.

Each of the examples laid out in this chapter can be explicitly linked with these components of the model. The network forms of these groups, like the structure of the internet itself, can be explained as rhizomatic, netwar or SPIN. Global activist networks have many centres or hubs but, unlike their predecessors, those hubs are less likely to be defined around prominent leaders. Movement integration has shifted from ideological integration towards more personal and fluids forms of association based on weak ties and informal connective structures. This leads to a rhizomatic political style, which fully realizes the organizational and mobilizational potential of internet network structures .

The successful protest against internet regulation in Spain demonstrated the speed of diffusion of protest and the effect of the internet as an informational, mobilizational and organizational tactic. The virtual sit-in to demonstrate solidarity with the Zapatistas reflects an impact of the internet on the framing process, using symbolic interplay in cyberspace. In the Seattle protests, alternative media were used as part of new communication technologies affecting protest and lobbying in cyberspace and the capture of windows of political opportunity. Anti-globalization movements (Indymedia, anticorporate hacktivism, etc.) are also given fuller attention in the section 'Social movement theory' (pages 53–71) in Chapter 2.

Chinese dissidents have used email spamming and proxy servers to access blocked sites, sophisticated websites, email lists, bulletin board sites, file-trading and e-magazines to express their dissent online. Dissident use of the internet exhibits characteristics that are typical of NSM activity. Links between issues raised in this chapter and the proposed thoretical framework include: issues of media sensitivity and event density, a political opportunity structure opened by the internet to allow cyberdissidents to reach international public opinion (online dissent, activism and arrest is extensively reported by foreign media), the structure of the online dissident movement (it seems to have no central leadership and looks network-type in character, for example the internet was crucial for recruitment for the CDP party), the use of technologically enhanced tactics, opening up alternative information and coordination networks, collective identity (cyberdissidents increasingly show solidarity towards each other), and the problematic relationship with the state (there is a crackdown on dissidents by the state). Those using the internet against their governments seek power, participation and democracy, making demands that governments are not only unwilling to provide but, more importantly, which prompt counterstrategies to crack down on these cyberdissidents (see the following section, internet as a medium: media effects on policy). The internet is, therefore, a battleground for these opposing interests, and it remains to be seen whether it will develop into a powerful engine for democratization, or will fall under the pressure and regulation of authoritarian regimes.

The internet as a medium

a) Analysing discourses (representations of the world, constructions of social identities and social relations); b) Control of information, level of censorship, alternative sources; c) Political contest model among antagonists – the ability to initiate and control events, dominate political discourse, mobilize supporters (Wolfsfeld); d) Media effects on policy (strategic, tactical, representational).

The anti-WTO protesters were able to initiate a newsworthy event, putting their opponents on the defensive. Using the internet, they could send stories directly from the street for the whole world to see, rendering the information uncontrollable. Thirdly, they were able to mobilize support by promoting an alternative frame for the event.

Dissidents acting against governments were able to use a variety of internet-based techniques (email lists, email spamming, BBS, peer-to-peer and e-magazines) to spread alternative frames for events, and a possible alternative online democratic public sphere (see alternative sources). A discourse analysis of BBS messages, for instance those asking, 'Can you establish your own political party? Can you set up a free labor union? Can you freely publish your political ideas? Organize groups? Demonstrate? As long as you don't have those few things, you don't have anything',[3] shows the determination of dissidents to question the monopoly of the state, the censorship and the repressive structure of Chinese politics, whilst creating an alternative public sphere.

An example of dissidents' use of the internet is spamming e-magazines to an unprecedented number of people within China, a method which provides recipients with 'plausible deniability'. The VIP Reference magazine is reportedly sent to between 250,000 and 300,000 Chinese email addresses. Also, file-trading networks like Kazaa and Gnutella can help dissidents communicate, since they have no central source and would be harder to turn off. China has responded effectively to the dissident challenge by implementing a combination of reactive measures, including blocking websites and jailing activists and proactive policies, such as distributing propaganda online and proffering e-government services. Also, in the other countries mentioned it is crucial to analyse the effect of government control and censorship.

Using Cronauer's argument about the long-term effects and the interlinkage of groups using emailing lists, the effects of networking, recruiting and the development of strategies, it is evident that the mobilization structure is affected greatly by online efforts. Hacktivism and its dilemmas were expressed vividly during the RNC protests. Hacktivist attempts at entering the mainstream demonstrate a dilemma that deserves attention in any future discussion or analysis of future sociopolitical cyberconflicts. These attempts involve seeking funding from large foundations and using the internet as a resource, or using different fractions within the hacktivist movement who insist on using the internet as a weapon for financial disruption and virtual attacks. In terms of the media component the new technological elements of 24-hour coverage, mobile reporting and organizing in protests demonstrate the notion that citizen journalism or open source journalism has entered mainstream. An undeniable fact when considering consider the example of the 7/7 terrorist blasts in London, where 1,000 photos and 20 pieces of amateur video were sent into the BBC news website alone, and even featured in newspaper front pages (Twist 8 July 2005). This has potential effects on policy, as the immediate coverage can have all sorts of negative and positive effects on the coverage and outcome of the outcome of conflicts.

In relation to China and internet censorship issues it is obvious that most governments get negative points for the freedom of their citizen's access to the network. Bypassing censorship and using techniques to get banned information or to transmit forbidden information affects media coverage in all these countries and this again can affect policy. Interestingly, the key words or the themes banned in most authoritarian regimes, if analysed discursively, point to either desired banned topics and ideologies, such as democracy, participation, revolution, reform, etc., or very negative ones, such as massacre, or historical events of oppression, repression and conflict. Online efforts, such as pro-democracy, activist or anti-government websites point to the fact that people believe in the power of the medium enough to organize and run thousands of these sites. In many cases, they are able to initiate and control events, and mobilize and recruit others for their cause, as in the case of sites in the Islamic world, in China, in Latin America, activist sites for anti-globalization and single-issue protests and mobilizations both on national and international levels.

This framework for looking at sociopolitical cyberconflict includes the most basic aspects of the phenomenon and it is not claimed to be definitive. It is offered as an analytical tool for future students of political conflict on the internet and as a way of opening space for discussion. The next chapter looks at a different type of cyberconflict, focusing on ethnoreligious conflicts, using conflict theory as the theoretical standpoint.

5   Ethnoreligious cyberconflict

This chapter looks at ethnoreligious cyberconflicts, that is real-world conflicts with ethnoreligious characteristics which spill over into cyberspace, such as the Kosovo conflict, the the Israeli–Palestinian conflict, the Kashmir dispute, the Sino-American incident and al-Qaeda on the web (Karatzogianni 2004a). It also includes an assessment of ethnic conflicts and a discussion of whether the internet could be used as a tool of conflict resolution, using examples of particular efforts in the Israeli–Palestinian conflict. The main argument concerning this type of conflict is that the groups involved use the internet, not as a resource with which to reframe the issues, but rather as a weapon, employing a method analogous to stone-throwing. Lastly, the chapter ends by linking these examples of ethnoreligious cyberconflict with the proposed integrated theoretical framework.

Ethnoreligious cyberconflicts

Kosovo

On 20 October 1998, the Kosovo Information Centre (KIC), which supports the party of the ethnic Albanian leader, Dr Ibrahim Rugova, reported that hackers claiming to be members of the Serbian terrorist organization Crna Ruka (Black Hand) hacked its webpage. The hackers left an image of the Serbian nationalist symbol and wrote in Serbian and English: '[w]elcome to the Web page of the biggest liars and killers ... Brother Albanians, this coat of arms will be in your flag as long as you exist' (BBC Online 25 October 1998). After the pages were restored, the hackers came back with a message saying who they were and posting the slogan 'Long Live Serbia'. The same cyber-unit also attacked other Albanian sites and the largest Croatian daily English newspapers. Croatian hackers retaliated by crashing pages of the Serbian National University Library.

Later, in April 2000, more than 50 websites were taken over by Serb hackers. The visitors to high-profile websites such as those of Adidas and Manchester United were surprised to see the image of a double-headed eagle, with the slogan 'Kosovo is Serbia'. Also hacked were the Albanian site Kosovapress and the Albanian newspaper *Koha Ditore*, as well as the Serbian Ministry of Information and Bosnian, Croatian and Yugoslav sites. The Serbian Ministry of

Information said that sites of Yugoslav providers, political parties and firms were attacked in a synchronized manner (BBC Online 14 April 2000).

The technical director of WebDNS, a domain-monitoring company, also commented that the attack was part of a sustained campaign. The way he explained it was that the hackers replaced the contact details in Network Solutions, under which most of the sites were registered, by transferring the contact addresses to a Yugoslav site and then to an Albanian one. They probably tricked Network Solutions by sending an email requesting a change of address and pretending to be from the companies that were hacked.

During the same conflict, but involving different players, US defence officials said on 8 November 1999 that the Pentagon refrained from unleashing an all-out computer attack on Serbia, because they were uncertain about the legal implications of launching the world's first cyberwar. They claimed that, in theory, they had the capacity to hack Milosevic's bank account and plunder Serbia's financial system. Furthermore, in May 1999, the Pentagon issued a 50-page booklet of guidelines for waging cyberwar called 'Assessment of International Legal Issues in Information Operations', which the Defense Department had been preparing since the first cyber-offensive used in Haiti in 1994. The report considers that information operations would be legal 'weapons' in the traditional law of war, thus rendering cyberattacks on civilian targets like universities or financial infrastructure a war crime. John Arquilla, a cyberwar expert at the Naval Postgraduate School in California, and Martin Libicki, a researcher on cyberwarfare for the Rand Corporation think-tank in Washington, both think that the Pentagon spied on Serbian computers, but did not attack them for strategic reasons (Borger 9 November 1999).

Despite these instances of cyberwarfare, the real novelty of the use of the internet during the Kosovo conflict was the amazing amount of emails sent to news organizations from the people in the region. As Chris Nuttal comments, '[i]f someone could write a program quickly enough, it should be possible to collate them from all over the Net and automatically build up an interactive map of Yugoslavia linking to accounts of the bombings town by town' (16 April 1999).

The Kosovo conflict is considered the first major conflict where cyberwarfare was used. NATO admitted that its website was blocked due to a bombardment of automated requests for information. During the bombing campaign, NATO web servers were subjected to sustained attacks by what NATO suspected were hackers in the employ of the Yugoslav military (Messmer 5 April 1999). The attacks included 'ping saturation' (denial of service attack where a target computer is overwhelmed with ping requests) and email viruses which brought NATO servers to a halt for a number of days. The attacks caused serious disruptions in the communication infrastructure of the organization.

Israel–Palestine

The increasing importance of cyberconflict is indeed very evident when it reflects conflicts belonging to the real world. In October 2000, Israeli and

Palestinian hackers engaged in adversial hacking when the prolonged peace talks between the two parties broke down. Until the beginning of November 2000, groups supporting either side in the conflict limited their online activities to defacements and Denial of Service attacks against websites affiliated with the Palestinian movement or Israeli nationalists. One example arose when an Israeli flag, Hebrew text and a piano recording of 'Hatikva', the Israeli national anthem, appeared on the Hezbollah home page (Hockstader 27 October 2000). Also, Palestinian hackers created a website called Wizel.com – a host for a FloodNet attack, a type of attack which reloads a targeted web page several times, rendering the site inoperable. The reaction was a sustained counterattack from pro-Palestinian 'cybersoldiers' from the US. The websites of the Israeli Army, Foreign Ministry and Parliament, among others, were attacked. Targets also included financial institutions, as a result of which, e-commerce sites crashed and there was an economic impact reflected in the Israeli stock market (Hockstader 27 October 2000).

The situation escalated in the first days of November 2000, when an anti-Israeli hacker attacked the website of one of Washington's most powerful lobbying organizations, the American–Israeli Public Affairs Committee (Aipac). Fred Cohen, a computer-security professor, commented at the time: '[w]hen you talk about war, you are talking about turning off the constraints that hold back people. You have people who want to break into computers, and now they have an excuse – they can do it for a cause' (Lemos 6 November 2000). The hackers published critical emails downloaded from Aipac's own databases and credit card numbers and email addresses of Aipac members. After the FBI was informed, the members of the organization, including a Republican Senator, were advised to cancel their credit cards and monitor their accounts. The hackers wrote that 'the hack is to protest against the atrocities in Palestine by the barbaric Israeli soldiers and their constant support by the US government' (BBC Online 3 November 2000). Aipac spokesman Kenneth Bricker said at the time that the hackers downloaded credit card numbers and about 3,500 names and web addresses from people who had contracted Aipac's website. The broadest list of the organization's 55,000 members was stored in a separate computer system and was not compromised.

The Israelis were not slow to retaliate. According to MAGLAN, an Israeli information warfare research lab, an Israeli supporter, 'Polo0', posted Palestinian leaders' mobile phone numbers, as well as information about accessing the telephone and fax systems of the Palestinian Authority, plus 24 different websites, 15 IRC channels and an IRC server through which the Palestinian movement communicates. Analysis by iDefense, a security monitoring agency, considered a number of key players in the cyber conflict. On the Israeli side, the wizel.com creators, a.israforce.com, Smallmistake and Hizballah attacked Palestinian sites.

On the Palestinian side was Unity, a Muslim extremist group, one of the forerunners of what is referred to as e-jihad or cyberjihad. Unity attacked the Tel Aviv Stock Exchange. Later they announced that their strategy was four-phased.

Phase one included crashing official Israeli government sites, phase two hit the Bank of Israel and phase three targeted the Israel ISP infrastructure, Lucent and Golden Airlines, and an Israeli telecommunications provider. They also said that they would not realize phase four, the destruction of e-commerce sites, but added: '[w]e warn the Zionists and their supporters that any attempt to touch any Anti-Zionist site will be faced with phase four of the cyberwar – causing millions of dollars in transactions' Gentile (8 November 2000). Unity also claimed in an email in February 2001 to have successfully attacked AT&T in retaliation for the company doing business – providing back-up in case of emergency – with the Israeli Defence Force, claiming to have blocked the site for 72 hours in one particular hit (Galvin 20 February 2001).

What distinguishes this cyberconflict from past ones is that it moved beyond being a game controlled by a few highly specialized hackers into being a full-scale action involving thousands of Israeli and Arab youngsters sending racist and occasionally pornographic emails to their opponents, while circulating for their supporters a range of website addresses with simple instructions on how to crush the enemy's electronic fortresses (Hockstader 27 October 2000). One site offered a menu of targets to attack, including the sites of Hezbollah, the Palestinian National Authority, Hamas and a dozen others (Hockstader 27 October 2000). The site encouraged users to click on the targets they would like to disable and offered a set of simple instructions for executing the assault. The whole process did not take more than a minute or two, and generated multiple high-speed attacks. IDefense's director of intelligence production, Ben Venzke, thinks that the Palestinians won this particular battle in cyberspace, because according to him, people on the Palestinian side were trying to learn how to hack overnight, in order to join the effort (Hershman 29 June 2001).

The political crisis in the beginning of 2002 in the Middle East spawned another increase in defacement attacks on Israeli web servers. Israel was the victim of ten out of 15 significant web defacements in the Middle East over the first two weeks of April 2002, according to security consultancy mi2g. Mi2g reported Israeli websites with the 'il' domain were defaced 413 times in 2001 – up 220 per cent from the year before – and Israel has been the biggest victim of web defacements over the past three years, suffering 548 of the 1,295 attacks in the Middle East. The most active anti-Israeli hacker group claims to be Egyptian and started its activities just after 9/11 (Leyden 17 April 2004).

India–Pakistan

Sympathizers on both sides of the Kashmir conflict (in northwestern India) have used cyberattacks to disrupt each other's computer systems and to disseminate propaganda. One of the first moments of cyberwarfare in the region was reported on 16 October 1998 by the Indian news agency PTI. Suspected Pakistani intelligence operatives had hijacked the Indian Army's only website, 'Kashmir: A Paradise', which gives the Indian view on Kashmir. The site was set up a month earlier as counter-propaganda to the dozens of sites supporting

Muslim Kashmiris seeking independence. The hackers had posted information on alleged torture of Kashmiris by the Indian security forces. The attack occurred as India and Pakistan began talks in Islamabad in an effort to ease tensions. The Pakistani hackers dedicated the 'new' site to 'all the Kashmiri brothers who are suffering the brutal oppression of the Indian army' (BBC Online 16 December 1998). The photographs of the site were overwritten with the slogans: 'Stop the Indians' and 'Save Kashmir'. Pictures showing Kashmiris allegedly killed by Indian forces were posed under headings such as 'massacre', 'torture', 'extra-judicial execution' and 'the agony of crackdown'. A government statement said the hackers changed the site parameters to divert visitors to a different server.

Among the propaganda on the site, there was a guest-book where visitors could leave comments. Two typical responses from the opposing sides were: 'this website is very biased and very unfair to the Pakistani point of view. This is just a whole charade by the Indians and 80% of it is absolutely untrue!' (Pakistani), and 'a whole hearted salute for my brothers fighting for our country with a religious maverick enemy' (Indian) (Nutall 5 October 1998).

In March 2000, the cyberconflict escalated when a group of Pakistani hackers defaced 600 websites and temporarily took over government and private computer systems. The majority of the sites were hacked after the Pakistanis broke into IndiaLinks, India's largest internet service provider. The team responsible, the 'Muslim On Line Syndicate', were described by their spokesperson as a group of nine ranging from 16 to 24 years of age (Hopper 20 March 2000). Their spokesman also described their method of taking control of a server, then defacing the site, after they had no more use for the data or the server. Their message was: '[w]e hope to bring the Kashmir conflict to the world's attention ... We wish that our Muslim brothers will be given the right to choose, as was promised them half a century ago' (Hopper 20 March 2000).

The number of pro-Pakistani defacements of Indian websites increased dramatically between 1999 and 2001: 45 in 1999, 133 in 2000 and 275 by the end of August 2001 (*The Statesman* 21 August 2001). However, the assault on Pakistani sites has not been as successful. There were reports that Indian hackers have repeatedly tried to hack the Pakistani newspaper *Dawn*, without any result. Nevertheless, they have left messages to their Pakistani counterparts like 'keep your hands off Indian sites', have threatened 'breaking the Internet backbone' of Pakistan, and have claimed that 'India is the superpower of Information Technology' (Joseph 23 December 2000).

More recently, an email worm in March 2003 appeared to be yet another salvo in a year-long war between Indian and Pakistani hackers. According to the Indian Snakes, authors of the worm known as Yaha, it was written to retaliate against Pakistani hackers, who, the Snakes said, were defacing websites in India. Yaha variants have been around the net for over a year. It began its life as a standard mass-mailer worm. Political messages first appeared in its E variant, released in June 2002, which attempted to launch elementary Denial-of-Service attacks against the Pakistani government's primary website. Yaha.Q attempted to launch a Denial-of-Service attack against five Pakistani sites, changing user settings on infected machines and containing a number of messages directed at Pakistani hackers, other virus writers and an anti-virus researcher.

Virus writers often use their creations to make a point, but very few are overtly political. Another example of a politically motivated virus attack was the Lion worm of 2001, which was, according to its author, intended to chastise Japan over the issue of Japanese textbooks that implied that the Japanese occupation of China and Korea was justified and beneficial to the occupied countries (Delio 13 March 2003).

US–China

Another example of ethnoreligious cyberconflict is the Sino-American cyberconflict which emerged during an international diplomatic incident. A US spyplane made an emergency landing on Chinese soil on 1 April 2001, after colliding with a Chinese fighter jet over the South China Sea, in a collision which killed the Chinese pilot. After this incident, Chinese hackers vowed to attack US sites, a threat which led hackers in the US to retaliate. According to UK computer security firm Mi2g, the Honkers Union of China hacking groups defaced 80 websites and the Americans defaced more than 100 during April 2001 (Left 4 May 2001). China's remote sensing satellite ground station was overwritten with a picture of a mushroom cloud, while in the US, the White House historical association was plastered with Chinese flags as were the departments of Health, Navy, Labor and the House of Representatives' email servers.

On 9 May 2001, Chinese hackers boasted they had defaced 1,000 US websites, but called a truce to the conflict. A statement by the Honker Union of China said that, having attacked 1,000 sites, they had reached their goal, and that any attack from that point on had no connection with them. Their American counterparts broke into hundreds of Chinese sites, leaving messages such as, '[w]e will hate China forever and will hack its sites' (Globe Technology 10 May 2001). After a meeting online between the Honkers Union and the Chinese Red Guest Network Alliance, it was decided that their attack would last a week, ending on 7 May, the two-year anniversary of the bombing of the Chinese embassy. They decided to keep the destruction of business websites to a minimum and attack government websites instead. They said that the point of the attack was to encourage people in the US to protest against their government and demand peace between nations. One hacker said: 'the U.S. wants the world to go to war. All people cherish peace, but the mildew dog government of the U.S. wants war. We will attack to send a message to the people of the U.S., to tell them we are all one, but they must stop their government from destroying the world' (Delio 30 April 2001).

Attacks that were discussed on an Internet Relay Chat during their meeting included defacing websites, emailing viruses to US government employees and flooding computers with garbage data. A US hacker collective dubbed Project China left this message on a Chinese site: 'Get ready to meet a strike force with strength the world has never seen before! We are going for all-out cyberwarfare on your gov.cn boxed and every box that you fucks haven't secured!' (Left 4 May 2001). The Xinhua News Agency reported at the time that US hackers had defaced the websites of the provincial governments of Yichun, Xiajun and Beijing, the Deng Xiaoping Universities, as well as the Samsung and Daewoo Telecom sites in South Korea. A South Korean government security agency blamed the Sino-US cyberwar for the 164 cyberattacks on South Korean websites that had occurred during that time. Computer analysts said that American and Chinese hackers were using Korea to get into rival countries' computer systems without revealing their identities, because South Korea has extensive links with both countries.

Interestingly, the Chinese government has been quite open about its future strategic military objective. In the spring 2001 issue of *China Military Science Journal*, a member of the Chinese Committee of Science, Technology and Industry of the System Engineering Institute wrote:

> We are in the midst of a new technology in which electronic information technology is the control technology. The technology provides unprecedented applications for the development of new weaponry ... Military battles during the 21st century will unfold around the use of information for military and political goals.
>
> (Chepsiuk 23 August 2001)

This is hardly surprising. As we have seen in the section on pages 128–143 in Chapter 4, the Chinese government has placed heavy and detailed regulations on the internet, has arrested dozens people for their online activities, has closed down thousands of internet cafés and has blocked thousands of sites. Reportedly, elements of the government have also hacked sites belonging to dissident groups.

Al-Qaeda[1] on the web

> The size and the structure of the internet provides virtual sanctuary. The internet provides the glue that links groups that operate within the ancient modes of organic order – religious, tribal, etc. that from the backbone of the physical world sanctuary, with the modern world's operational environment. However, the internet is more than merely a communications medium, it is a place of sanctuary in itself.
>
> (E-Resistance@yahoogroups 20 August 2004)

Most militant groups now rely on the web to recruit new adherents. Terrorist groups rely increasingly on internet chat rooms, more anonymous than traditional websites. Gabriel Weiman, a professor at University in Haifa in Israel, began tracking terrorist-related websites eight years ago, and found 12; today there are more than 4,500 (Coll and Glasser 7 August 2005).

The web's growing centrality in al-Qaeda-related operations has led such analysts as former CIA director John E. McLaughlin to describe the movement as primarily driven today by 'ideology and the Internet' (Coll and Glasser 7 August 2005). The web's shapeless disregard for national boundaries and ethnic markers fits exactly with bin Laden's original vision for al-Qaeda, which he founded to stimulate revolt among the worldwide Muslim ummah, or community of believers.

It was said that the Hamburg cell around Muhammad Atta, one of the presumed suicide pilots, was systematically using the internet to organize the 9/11 attacks (Aliefudien 26 December 2004). For security reasons, these individuals rarely used their private PCs from home, but more often visited internet cafés. The US intelligence agencies came to this conclusion when they arrested the alleged mastermind of the attacks, Abu Zubayda, and put his personal computer through the hoop. In it, they found a large number of encoded messages – the last ones sent on 9 September. Vince Cannistraro, the former head of the CIA's counter-intelligence unit once said: 'Internet communications have become the main communications system among al-Qaeda around the world because it's safer, easier and more anonymous if they take the right precautions, and I think they're doing that' (Aliefudien 26 December 2004).

In November 2001, as the Taliban collapsed and al-Qaeda lost its Afghan sanctuary, Osama bin Laden biographer Hamid Mir watched 'every second al-Qaeda member carrying a laptop computer along with Kalashnikov' as they prepared to scatter into hiding and exile (Coll and Glasser 7 August 2005). On the screens were photographs of September 11 hijacker Mohamed Atta. With laptops and DVDs, in secret hideouts and neighbourhood internet cafés, young coding jihadis have sought to replicate the training, communication, planning and preaching facilities they lost in Afghanistan with countless new locations on the internet. Al-Qaeda suicide bombers and ambush units in Iraq routinely depend on the web for training and tactical support, relying on the internet's anonymity and flexibility to operate with near impunity in cyberspace. In Qatar, Egypt and Europe, cells affiliated with al-Qaeda that have recently carried out or seriously planned bombings have relied heavily on the internet. Al-Qaeda's innovation on the web 'erodes the ability of our security services to hit them when they are most vulnerable, when they are moving', as Michael Sheuer, former CIA chief of the unit that tracked bin Laden explains. 'It used to be they had to go to Sudan, they had to go to Yemen, they had to go to Afghanistan to train. An al-Qaeda operative no longer has to carry anything that is incriminating. He doesn't need his schematics, he doesn't need his blueprints, he doesn't need anything that is incriminating'. Everything is posted on the web or 'can be sent ahead by encrypted Internet, and it gets lost in the billions of messages that are out there' (Coll and Glasser 7 August 2005).

Al-Qaeda militants have defied a crackdown and the loss of senior leaders in Saudi Arabia by using the internet to win over new recruits in Osama bin Laden's birthplace (Reuters 16 November 2004). Despite the killing of top contributors, including one of its leading web magazine editors Issa Saad bin Oshan, the group has continued to publish its two widely distribute magazines regularly for the past year. Paul Eedle, a London-based analyst who closely follows al-Qaeda sites, explains it as follows: 'This shows how a small group can continue a campaign using the Internet. Before the days of the Internet a group would pretty much fade from view if they were reduced in numbers like al-Qaeda in Saudi Arabia' (Reuters 16 November 2004).

After an attack on a hotel in central Baghdad, the group 'al-Qaeda in Mesopotamia' released an internet statement claiming credit, while one group even released its own videotape of the bombing, along with statements explaining why and how it chose that target. Within hours, all of it was appearing not only in Arabic websites and chat rooms, but also on television stations and even in some Western news reports (Worth 13 March 2005).

It is an all too familiar ritual. Hours after an attack on an American convoy or an Iraqi police patrol, a brief statement begins appearing on Islamist websites claiming it was carried out by fighters loyal to Abu Musab al-Zarqawi (Worth 13 March 2005). Everyday messages appear on the web offering encouragement to resistance fighters. Zarqawi's group started an internet magazine, complete with photographs and 43 pages of text. According to Michael Doran, a professor of Near East Studies at Princeton University who monitors traffic on Islamist web sites and chat rooms, the magazine is partly a reaction against the Arab state media, which often misrepresent terrorist attacks according to any other Qaida-linked web publications. Other Islamist groups are joining the effort, including one calling itself the Jihadist Information Brigade.

Maaskar al-Battar (al-Bataar Training Camp), a jihadi online publication which has ceased publication, with motivational material interwoven with military training input, aimed to build a jihad culture among Muslim youth and to equip them to fight against those who have 'invaded the Islamic world'. With Maaskar al-Battar, training to become a jihadi was available at the click of a mouse. The first issue drew attention to the convenience of online training it provided. 'Oh Mujahid (holy warrior) brother, in order to join the great training camps you don't have to travel to other lands ... Alone, in your home or with a group of your brothers, you too can begin to execute the new program. You can all join the al-Battar Training Camp' (Ramachandran 31 May 2005). Al-Khansa, a magazine aimed exclusively at women and named after a woman poet who was a close associate of the Prophet Mohammed, was designed to motivate women to participate more actively in the jihad. Jihadi online publications are not just about getting matter, putting together content and posting it on the internet. They are also concerned with survival, which is determined not so much by funding and circulation as by evading counter-terror experts. They constantly move URLs and change addresses regularly, sometimes embedding themselves within other websites to evade detection. Another jihadi online publication named Dhurwat al-Sanam, Arabic for the 'highest or more virtuous belief/insight' has been posted by the official spokesman of al-Qaeda in Iraq (Ramachandran 31 May 2005).

Imam Samudra, charged with engineering the devastating Bali nightclub bombings, published a jailhouse autobiography justifying the Bali attacks, which killed 202 people. This was not surprising. What was a surprise was the chapter at the back of the book entitled 'Hacking, Why Not?' There, Samudra urges fellow Muslim radicals to take the holy war into cyberspace by attacking US computers, with the particular aim of committing credit card fraud, called 'carding'. The chapter then provides an outline on how to get started. Samudra is among the most technologically savvy members of Jamaah Islamiah, an underground Islamic radical movement in Southeast Asia that is linked to al-Qaeda. He sought to fund the Bali attacks in part through online credit card fraud, according to Indonesian police. 'This is hacking for dummies. But in this day and age, you don't have to be an expert hacker to have a tremendous impact', comments Evan F. Kohlman, a US consultant on international terrorism who reviewed the chapter. Kohlman and other cyberterrorism experts say the kind of fraud preached by Samudra is becoming increasingly attractive as a source of funding for al-Qaeda operatives in several regions of the world (Sipress 14 December 2004).

Three internet cafés, including one located in downtown Riyadh, have been raided by security officials in a move to track down terrorists who have been using public cafés to exchange information, post terror messages and issue threats to organizations, government agencies and nations (Khan 18 January 2005). Arif Ziauddin, manager of one of the raided cafés, stated that many terror suspects have turned to cyberspce to communicate with their accomplices since the 12 May 2003 bombings in Riyadh.

On 1 June 2004, al-Qaeda terrorists conducted one of their most spectacular operations – a brutal assault on the Saudi oil town of Khobar, replete with seek-and-destroy missions targeting non-Muslims and gun battles with security forces (Kimmage 16 June 2004). An internet post by a man who calls himself Fawaz Bin Muhammad al-Nashmi, to a forum called Al-Qal'ah (the Fortress), recounts a heroic 'battle' in which he and comrades set out to cleanse the Arabian peninsula of infidels. The posting resembles a Hollywood action film. Although al-Qaeda adherents are commonly described as having a medieval world view, their rhetoric and self-image owe as much to blockbuster movies and Mortal Kombat as to epic tales of seventh-century Islam. Al-Nashmi uses the word 'ilj', or 'unbeliever', to refer to the non-Muslism he and his comrades murder. It has become a jocular pejorative for US soldiers, and Al-Nashmi uses it interchange-ably with 'cafir', the standard put-down for an infidel. As improbable as it is that these events occurred as he describes them, his idealized narrative presents them as they appear to the mind's eye of al-Qaeda. The pool of potential recruits teems with young men adrift amid feelings of humiliation and powerlessness, eager for a world view that answers their questions, and hungry for action. In his account of a blood-soaked day, al-Nashmi gives them what they lack – power over life and death, a mission to rid the world of enemies and violence as the path to deliverance.

Efforts abound at undermining al-Qaeda's virtual operations, networking, mobilization, propaganda, recruiting and fundraising. In fact, various research centres, universities and think-tanks have been funded, or are preparing proposals for funding, on several counterterrorism projects. For instance, documents obtained by the Electronic Privacy Information Centre through a Freedom of Information Act show that the US Central Intelligence Agency and the National Science Foundation collaborated to fund researchers developing software to electronically spy on internet chat rooms. The project is described as the 'fully automated surveillance system for data collection and analysis in internet chat rooms to discover hidden groups' (Indymedia 15 December 2004).

A new generation of software called Starlight 3.0, developed for the Department of Homeland Security by the Pacific Northwest National Laboratory (PNNL), can unravel the complex relationships between people, places and events (Gartner 10 May 2005). Anticipating terrorist activity requires continually decoding the meaning behind countless emails, web pages, financial transactions and other documents, according to Jim Thomas, director of the National Visualization and Analytics Centre (NVAC) in Richland, Washington. In September 2005, NVAC, a division of the PNNL, will release its Starlight 3.0 visual analytics software, which graphically displays the relationships and interactions between documents containing text, images, audio and video. Starlight quadruples the number of documents that can be analysed at one time – from the previous 10,000 to 40,000 – depending on the type of files. It also permits visualizations to be opened simultaneously, which allows officers for the first time to analyse geospatial data within the program.

Moreover, British chief police officers are asking the UK government for new powers that would allow them to attack terrorist websites (Illet 25 July 2005). Ken Jones, chairman of the ACPO Terrorism and Allied Matters Committee, said that the evolving nature of the current threat from international terrorism demands that those charged with countering the threat have the tools they need to do the job. But a representative of Spy.org.uk, a civil liberties advocacy website, in an email to silicon.com, wrote: 'Who exactly is going to define what a terrorist website is? There are none of these hosted in the UK, so the targets must be abroad . . . The only people who seem to have a legal hacking law at the moment are the Australians, but it does not appear that they have dared use it against overseas targets.'

Other legal issues are also controversial. For example, the case of Babar Ahmad, who after being arrested under Britain's draconian and catch-all Terrorism Act, and released without charge, presumably for lack of actual evidence, has been arrested in London in order to be extradited to the USA. He is accused of running a website in the USA, which apparently supported the Taliban in Afghanistan and some rebels in Chechnya in Russia. Neither of these were at the time, or are now, proscribed terrorist organizations in the UK, yet this British citizen is in the process of being extradited to a foreign country – the USA – regarding possible actions in third and fourth countries – Afghanistan and Russia (E-resistance@yahoogroups 20 August 2004).

Ethnic cyberconflicts

Brief discussions have thus far been made of the Israeli–Palestinian, Indian–Pakistani and US–Chinese cyberconflicts, which are considered to be ethnoreligious cyberconflicts. The next section looks at various cyberconflicts that are more ethnic in character, namely China–Taiwan, China–Japan and Colombia (more of an internal conflict), in order to include empirical evidence and discussion of different types of cyberconflict.

After Taiwanese president Teng-Hui talked about a two-state theory to describe the relationship between China and Taiwan in what was perceived by many as a declaration of independence, Chinese hackers attacked many of the Taiwanese government's websites. The hackers during the first week of August 1999 filled the pages with political slogans and warnings to the 'separationists'. On the rebound, Taiwanese hackers attacked Chinese government sites, with successful hack attempts. The interesting part was that Taiwanese hackers added many liberal elements to their hacking such as music (pop songs and the national anthem), animations and comments citing popular culture.

The Taiwanese authorities warned that hacking even a Chinese site is illegal and that hackers would be prosecuted. The response from a representative of Taiwanese hackers was that their movement was an autonomous action against any kind of information warfare invasion, and a dismissal of the government's countermeasures as unethical (Jame 10 August 1999).

Also, in February 2000, Chinese hackers launched an attack against major Japanese companies. The reason for this was what they perceived as Japan's hard line against China. They left messages such as 'Down with Japanese militarism' and 'Kill all Japs' (Jame 10 August 1999). Japan more recently has been targeted twice. During the first week of April 2001, pro-Korean hackers attacked Japanese organizations responsible for the approval of a new history book. The textbook glossed over atrocities committed by Japan during World War II and the occupation of China and South Korea. The attackers were mainly Korean students, who crashed several websites belonging to Japanese organizations, including Japan's education ministry, the Liberal Democratic Party and the publishers of the book. Then in August 2001, pro-Chinese hackers defaced several sites belonging to Japanese companies after Japan's Prime Minister visited a controversial war memorial, the Yasukuni Shrine (Agence France Presse 14 August 2001). Chinese hackers have demonstrated their willingness to use cyberspace as a platform for protests and patriotic nationalism – though this is not to say that other nationalities have not been doing the same. Similarly, in Malaysia between April 1999 and April 2000, there were 89 hacker attacks on 60 government agencies, including sensitive data-carrying targets like the treasury, public works, social security and immigration, along with major non-government sites such as Malaysian Airlines.

Japan has bolstered the defence of its computer systems in the face of a surge in cyberattacks believed linked to anti-Japanese sentiment in Asia during the spring of 2005. Government officials were reluctant to publicly pin the attacks on Chinese and South Korean hackers because of the difficulty of identifying their source, but the attacks coincided with violent anti-Japanese protests in China. This actually fuels the argument that 'real' conflicts have the tendency to transfer online to the extent that cyberconflicts and hackattacks can be used as a barometer for real life conflicts. Cyberattacks hit Japan's National Police Agency, Self-Defence Forces and the Defence and Foreign Ministries, other businesses and a Tokyo war shrine criticized in Asia for honouring convicted war criminals. An official in Tokyo's Yasukuni Shrine, which honours war criminals among Japan's 2.5 million war dead, said, on condition of anonymity, that the site was hit with 15,000 attacks per second (Lateline News 11 May 2005). Experts say the recent flow of internet intrusions marks a shift from 'kids' play' hacking to organized, full-scale, politically motivated assaults. 'People have discovered that they can conduct "digital demonstrations" at any time. It's almost unthinkable that a huge flock of Chinese people could come to Japan and hold a protest. But digitally is possible', comments Itsuro Nishimoto, of SecureNet service, a division of Internet security LAC (Lateline News 11 May 2005).

Colombia

After the Revolutionary Armed Forces of Colombia (FARC) hijacked a commercial plane and kidnapped a senator at the end of February 2002, the government called off peace talks and decided to send troops to reclaim a Switzerland-sized territory it conceded to the rebels in 1998. While there were battles on the ground, another crucial battle was being fought in cyberspace to influence public opinion.

On their multi-lingual website, FARC condemn the government's move, rally their foreign supporters and swear to fight to the finish. FARC commandos wrote, in a press release published on their site the day after President Andres Pastrana ended negotiations in a televised speech: '[o]nce more the Colombian oligarchy impedes the dialogue that would allow the structural, economic, political, social and military changes that Colombia requires to escape the profound crisis that liberal and conservative governments have historically mired in' (Scheeres 4 March 2002). They continued in an online editorial: '[w]hile there is no definite solution to the great injustices suffered by our people, while resources are spent on war and not peace, we will continue forward in the conquest of power for the construction of the New Colombia' (Scheeres 4 March 2002).

2005 was the first year in more than a decade in which no Colombian reporters were killed. However, the statistic, while welcome, was due more to the increased caution being practised by journalists rather than an improvement in the country's situation. From a journalistic perspective, this is cause for concern, because it appears to have resulted in a self-censorship that has rendered serious investigative journalism virtually non-existent, and led to a distorted portrayal of Colombia's conflict.

Reporters are fearful of the reactions of right-wing paramilitaries and leftist guerrillas to the stories they report. Paramilitaries from the United Self-Defence Forces of Colombia (AUC) are prominent in Puerto Asis and other towns in Putumayo, while the FARC guerrillas control many of the rural areas. Due to their fear of the armed groups, local journalists in Putumayo practise what they call 'social journalism', in which they focus on social and political stories that do not involve the armed groups (Leech 14 January 2005).

Is the internet a tool for conflict resolution or an escalating factor?

The examples of ethnoreligious and sociopolitical cyberconflict show that cyber-attacks follow actual conflict, and are therefore closely linked with the real world. Also, cyberspace can be used as a battleground for future warfare, as some experts argue. On the other hand, in the pursuit of potential positive uses of the internet as a tool for communication and conflict resolution between opposing sides in a conflict, the Palestine–Israeli conflict emerges as a constructive example.

One of the initiatives taken to promote cooperation in the Middle East region through the internet is the Middle East Virtual Community (MEViC). MEViC (www.mevic.org) seeks to create a virtual community of academics and intellectuals in order to foster cooperation, collaboration in research, communication and understanding in the region. The organization currently has participants from Palestine, Egypt, Jordan, Israel, the Gulf, Turkey and Morocco. Working as individuals, they have been able to bypass political and institutional constraints that have been characteristic of the region. As Dachan puts it, 'the singular aspects of computer mediated communication are particularly relevant to creating the kind of community and discussion groups dealing with the issues related to regional conflict, and should serve as an important component of community based diplomacy'. Moreover, such a community has the potential to create strong dialogue, especially when there is a group of committed individuals, limited by neither time nor space, who have the capacity to plan and organize, to access regional research centres and to use online document delivery systems.

Another interesting attempt at dialogue is an internet community called the Middleast Abrahamic Forum (MEAF). On the homepage of the community, it is mentioned that the forum is devoted to interfaith dialogue between the three Abrahamic traditions, Islam, Christianity and Judaism. Its founder and manager, Mohamed Mosaad, has described the first seven months of the life of this community. The values emphasized by the majority of members included stressing the importance of unity between the different religions. The forum participants exchanged ideas reflecting the beliefs common to the different religions. Then they moved to discussing the ethics and value of dialogue itself and its ethics, and while they became more convinced of the differences between them, they remained committed to a search for peaceful coexistence. Mosaad writes that, although there were threads calling for a debate to prove which religion is the right one, or reminding the dreamy members of the conflictual aspects of reality, participants in the forum still managed to develop a common language (Mosaad undated).

Also, Mosaad is interested in the individual and collective identities of the MEAF virtual community. Responding to a thread raising the question of the group's identity, the members defined themselves in terms of nationality, ethnicity, religion, gender, age and political position. Some shaped their identity in opposition to an 'Other'. Some used the term 'I', others used 'we'; some were talking as members of a community, while others were representing different identities from outside. It is useful to include here Mosaad's comment that identity becomes an issue when one's status quo is threatened. He argues that, in this particular community, people felt threatened in two situations. First, they felt threatened by a political situation or international media coverage from outside the community, which operated in a way which threatened their identity (e.g. defending Muslims from being called terrorists after a suicide bombing). Second, they became defensive when their identity was threatened by an inside alternative. For instance, when a Jewish member said that Jerusalem should belong to the three traditions, some Jewish members stressed their own Israeli identity and their critic's American identity (Mosaad undated).

As far as their collective identity is concerned, Mosaad explains that, when receiving a new member, the old members replied in a sense of 'we welcome you'. When dealing with a rebelling member, the rest of the members showed solidarity against her. It did not matter to which country or religion he or she belonged. Every member responded in the same way to refute her, and when the member could not be convinced politely, they would start to attack the opponent in his thread, showing solidarity with their community. Even more interestingly (since, as Mosaad notes, a collective identity is not simply the drawing of a cognitive boundary; it is also an emotion), many of the members formed good friendships via phone calls and emails. As is common in online communities, when one member noticed the absence of another active member, he would send messages to check on him.

The power game turned out to be a game for information. Social movements, instead of exercising a physical action, acted to control the way information would be interpreted. That is why communities like these are becoming important as initiatives in regional conflicts. An example was when the MEAF community participated in a campaign to stop the demolition of Palestinian homes, where members of the community bombarded the Israeli PM, the general secretary of the UN and the then US Secretary of State Madeline Albright.

A third effort in the same region to build communication bridges through the internet was initiated by the Project for Arab–Jewish Dialogue of the Program in Conflict Resolution at Bar-Ilan University, with Palestinians counterparts from the Bethlehem-Hebron area. Participants from Bar-Ilan University and Beit Ommar village participated in an experimental email dialogue for approximately two months during the late spring of 1998 (Mollov *et al.*). Eight participants on each side were paired up and given instructions to introduce themselves to each other and describe a Muslim or Jewish holiday and conduct clarifications with each other on this topic. The organizing parties had to overcome major technical difficulties, namely, installing the first email location in Beit Ommar. Due to the success of this first dialogue, a second one was attempted, this time focusing directly on two holidays, Ramadan and Rosh Hashanah. The email effort was meant to explore possibilities for precise data transmission and comprehension across the Israeli–Palestinian divide, as well as to facilitate the positive relationship-building necessary for both conflict resolution and effective coactivity. Eight participants from both sides took short examinations on the two holidays before and after the email exchange, and concluded the two-month cycle with actual face-to-face workshops. An example of their email messages follows.

An Israeli student wrote:

> I learned that Ramadan is one of the five foundations of Islam, and that every Moslem has to fast for the whole month of Ramadan, in which the Quran was given to Mohammad on the night of the 27th. I didn't quite understand what do you do during the days of the fasting, are there any special thing you do during the day? Special prayers? Special places to go to? Another question: Do all Moslems fast on this month or just the very religious ones ...
>
> (Mollov *et al.*)

His Palestinian partner responded:

> It is so nice to hear from you, and it was a nice time that we spent at Bar-Ilan University. Regarding your questions: we spend time during the day in Ramadan as usual, we go to work normally, and we pray the five prayers but we add in the last prayer an extra prayer and during these prayers we read the whole Quran during the month ... Not all Moslems fast, but we can say most of them. I would like to ask what does Rosh Hashanah symbolize? And how many days do you celebrate?
>
> (Mollov *et al.*)

During the two-month email cycle, in total 65 messages were exchanged and the organizers were impressed by the richness of the message content and the social information exchanged. Furthermore, while both groups performed poorly in their exams prior to the dialogue, with the Israelis scoring 58 per cent and the Palestinians 51 per cent, after the dialogue the Palestinians increased their knowledge to 73 per cent and the Israelis to 61 per cent. Palestinian results can be partly explained due to the fact that Israeli communication styles place less emphasis on social interaction and speak directly to the point – that is to say, Arab participants acquired greater amounts of information from their Israeli partners than vice versa (Mollov *et al.*).

Despite these efforts at reconciliation, the cyberbattle continues for hearts and minds in the Israeli–Palestinian conflict. Examples include the pro-Palestine Electronic Intifada (www.electronicintifada.com), a site that encourages media activism and features daily reports from the Palestinian territories, while in the pro-Israel camp there is the New York-based Israel Support Group (ISG) (www.israelsg.com), which hosts a comprehensive site with news, video reports and activist guides (*Foreign Policy*, July/August 2002). One of the largest Islamic websites, IslamiCity (www.islamicity.com), reaches about 50 million people a month and features polls, TV and radio broadcasts, and religious guidance. Similarly, Middle East news online (www.middleeastwire.com) partners with 120 content providers worldwide and uses a network of reporters to disseminate information about the entire region. 'The impact of the Internet on the media is that they are hearing from people on both sides of the conflict', comments Rania Awwad, a representative for Palestine Media Watch (www.pmwatch.org) (*Foreign Policy*, July/August 2002). Also, the Middle East Media Research Institute (MEMRI) (www.memri.org) uses the web to monitor Arab media for anti-Israel content. All this cyberactivism means that journalists are subject not just to increased scrutiny, but also to massive lobbying campaigns.

Some observers worry that the proliferation of independent websites, media monitors and lobbying campaigns will have a polarizing effect on the conflict. 'Muslims get news from a Muslim perspective. Jews get news from their perspective. There is a gap in understanding what is happening in the conflict', says IslamiCity's Mohammed Abdul Aleem (*Foreign Policy*, July/August 2002).

Conclusion

In ethnoreligious cyberconflicts, when an anti-Israeli activist attacks the Aipac site, an Israeli sets up a website of targets to attack. When Pakistanis attack India's largest internet service or when Chinese and American nationalists hack each other, the use of the internet is different from that found in sociopolitical cyberconflicts. The parties in ethnoreligious cyberconflicts do not usually use the internet in order to mobilize support or influence public opinion. They use it as a weapon.

Cyberconflict should not be dismissed as just a series of catfights between computer geeks. Mi2g chief executive D.K. Matai has argued that cyberwarfare could be used as a barometer for political tensions around the world. 'The tense situation in the Middle East is reflected in both covert and overt hack attacks' (BBC Online 16 April 2002). In addition, cyberconflict is a phenomenon that includes a variety of actors with different characteristics, many of whom cannot be easily distinguished as either terrorist or activist in nature. Accordingly, the political game between parties in an ethnoreligious conflict or among social activists engaged in a sociopolitical one is neither clear nor fully developed at this early stage.

Nonetheless, it is important to emphasize that two different kinds of cyber-conflict are being discussed, and that they should be treated as distinct. The groups in sociopolitical cyberconflicts initiate a newsworthy event by putting the other side on the defensive, sending stories for the whole world to see, rendering information uncontrollable or mobilizing support by promoting an alternative frame for the event. On the other hand, in ethnoreligious cyberconflicts, opposing parties tend to use the internet as a weapon. While they might initiate events, they fail to promote an alternative frame. They simply attack the enemy or defend their electronic territory.

This means that cyberconflict needs to be monitored, and the actors and practices involved in such conflict need to be researched and evaluated. It is important that to increase understanding of the nature and trajectories of conflict with an internet component, for a number of political reasons. These include improving conflict resolution and mediation strategies, and developing ways of bringing cyberwar activities under the purview of international agreements governing conflict.

This section looked at ethnoreligious and ethnic cyberconflicts, in particular cyberconflicts between Israelis and Palestinians, Indians and Pakistanis, and Americans and Chinese, as well as incidents in Southeast Asia, Colombia and Kosovo, and attempts at conflict resolution using the internet as the platform.

The integrated theoretical framework proposed in the section on pages 86–93 in Chapter 2 encourages a focus on the following parameters when looking at ethnoreligious cyberconflicts.

Ethnoreligious cyberconflicts

a) Ethnic/religious affiliation, chauvinism, national identity; b) discourses of inclusion and exclusion; c) information warfare, the use of the internet as a weapon, propaganda and mobilizational tool; d) conflict resolution depends on legal, organizational framework, number of parties issues, distribution of power, values and beliefs.

In the Israeli–Palestinian cyberconflict, we see the use of national symbols when hacking websites, such as the Israeli flag, Hebrew text and even a recording of the Israeli national anthem on the Hezbollah home page. This explicitly urges us to look at issues of national identity, nationalism and ethnicity. Secondly, the language used by hackers, as in the AIPAC hack ('the hack is to protest against atrocities in Palestine by the barbaric Israeli soldiers and their constant support by the US government'), relies on an 'us' and 'them' mentality, where Israelis and their American supporters are portrayed as barbaric, reflecting discourses of inclusion and exclusion. Thirdly, the internet in ethnoreligious cyberconflict became a battleground and was used as a weapon by both sides, and full-scale action by thousands of Israeli and Palestinian youngsters involved both racist emails and circulating instructions on how to crush the enemy's website.

Similarly, in the Indian–Pakistani cyberconflict, the Indian army's website was set up as a propaganda tool, and hacked pictures of alleged tortures of Kashmiris by Indians were placed on the site, in a similar propaganda tactic. Also, the internet was used as a weapon, when the worm Yaha was released by Indian hackers. In terms of national identity and discourses of exclusion and inclusion, the following is telling: 'a whole hearted salute for my brothers fighting for our country with a religious maverick enemy' and '[w]e wish that our Muslim brothers given the right to choose, as was promised them . . .' In both sentences, religion is mentioned (religious affiliation), the word 'brothers' (collective identity and solidarity), and 'our country' and a promised land.

In the rest of the examples, similar links can be made with the proposed framework. Again this is clear in the discourse used: 'We will hate China forever and will hack its sites' (US–China); 'Down with Japanese militarism, Kill all Japs' (China–Japan); 'Welcome to the web page of the biggest liars and killers, Long Live Serbia' (Kosovo–Serbia). In all of these cases, a nation is identified as the enemy by nationality and appropriate adjectives are used to create the inclusion/exclusion effect.

The conflict resolution component in the framework is included due to the attempts at such resolution through the internet. The three examples (MEViC, MEAF and the Project for Arab–Jewish Dialogue) place emphasis on community-based diplomacy, interfaith dialogue (which reflects the importance of exchanging ideas reflecting the common beliefs between different religions) and building a positive relationship as a platform for resolution.

The internet as a medium

a) Analysing discourses (representations of the world, constructions of social identities and social relations); b) control of information, level of censorship, alternative sources; c) political contest model among antagonists: the ability to initiate and control events, dominate political discourse, mobilize supporters (Wolfsfeld); d) media effects on policy (strategic, tactical, representational).

In Colombia and Kosovo, the internet was used as a medium for disseminating information to the extent that, in the latter conflict, a commentator said that if one was to collate the amount of emails sent to news organizations, one would build up a interactive map of Yugoslavia linked by accounts of bombings town-by-town. Online news organizations like the Electronic Intifada and the Israeli Support Group also represent alternative sources for the explanation of conflict, can be used for propaganda and mobilization, and reflect a political contest between Israelis and Palestinians to dominate political discourse.

The al-Qaeda network and its ideology relies more on common religious affiliation and kinship networks than strict national identity, which fits well with the borderless and network character of the internet. Secondly, the internet has been used as a primary mobilizational tool, before 9/11, especially after the breakdown of cells in Afghanistan, Saudi Arabia and Pakistan. On the internet, al-Qaeda is replicating recruitment and training techniques and evading security services, because they cannot be physically intercepted, due to the virtuality of their networks. The internet is used as a propaganda tool via electronic magazines, training manuals and general recruitment sites, as well as a weapon for financial disruptions aiming at financing operations, or stealing data and blueprints.

On the organizational aspect and in order to avoid confusion, it is crucial to note that al-Qaeda is a unique case, because as an ethnoreligious actor in a cyberconflict, although expected to be more hierarchical – like the hackers in Israeli–Palestinian cyberconflict who use the internet more as a weapon than a resource – it is instead networked, in contrast to the rest of the ethnoreligious cyberconflicts, and uses the internet in all sorts of resourceful ways. A possible explanation for this is twofold: firstly, their ideology does not rest on national identity, so it is open to international networking through religion and kinship/family; and secondly, they are not left with any other choice but to use the internet in resourceful ways, because they are not legitimate and they are not going to be given a chance at conflict resolution.

This chapter has attempted to describe some cases of ethnoreligious cyberconflict, to understand their operation, to examine the current methods and weapons used and link these examples with the proposed integrated theoretical framework. The conclusion is that the knife can cut both ways. On the one hand, evidence suggests that cyberconflicts can spill over to the real world with information warfare becoming a threat to governments and businesses. On the other hand, the inexpensive, easy-to-use and interactive style of the medium can provide significant assistance in bringing opposing sides together who would otherwise not communicate due to spatial, political, religious or other differences. The next chapter examines the effect of the internet on the March 2003 Iraq conflict.

6 The effects of the internet on the 2003 Iraq war

> In the particular case of post-September 11 information warfare, the assertions of US unity by the Bush administration have resulted in a context in which public statements directed to the international community are interpreted as representative of the US as a political entity and not just the utterances of a particular individual in the current administration.
>
> (Walls 2002: 119–127)

Before and during the March 2003 Iraq war, information technologies, and particularly the internet, inspired several groups belonging to all sorts of different backgrounds and ideologies to voice their opinion on the war and, in certain instances, to engage in symbolic hacking against opposing groups or institutions. But still, the principal novelty of this conflict was the effect of the internet on war coverage. This section looks at three levels of the internet's role in the conflict: its effect on the organization and spread of the peace movement, its impact on war coverage and the issue of war-related cyberconflicts.

Before delving deeper, it would be helpful to explain again that, with the advent of the internet, new forms of conflict have emerged. These forms are not directly linked with information warfare but, rather, connect to a more subtle form of societal netwar (Arquilla and Ronfeldt 2001), where new social movements, ethnic groups and terrorists use the internet to organize, acquire resources and attack 'the other side'. Despite the high-tech name, the groups involved have quite traditional political goals – power, participation, democracy, alternative ideologies – using, however, a postmodern, interactive medium.

There are two types of cyberconflict (CC). The groups in sociopolitical cyberconflicts (for example, the peace and anti-globalization movements) initiate a newsworthy event by putting the other side on the defensive, sending stories for the whole world to see, rendering information uncontrollable and mobilizing support by promoting an alternative frame for the event. On the other hand, in ethnoreligious cyberconflicts (such as Israeli–Palestinian hacking and Indian–Pakistani activity), opposing parties tend to use the internet as a weapon. While they might initiate events, they fail to promote an alternative frame. They simply attack the enemy or defend their electronic territory (Karatzogianni 2004a: 46–55). Thereby, the different kinds of cyberconflict will be analysed in the context of international conflict theory for ethnoreligious cyberconflict, and social movement theory for sociopolitical cyberconflict, while keeping in mind that this takes place in a new media environment, using media theory.

The most recent opportunity to apply this theoretical framework arose with the recent war in Iraq. There, the distinction (ethnoreligious/sociopolitical) worked quite well (Karatzogiani 2004b). There was sociopolitical CC before the war, starting with the peace groups organizing demonstrations and events through the internet; and, while the tension was mounting, there were hackings between anti-war and pro-war hacktivists (sociopolitical), but also between pro-Islamic and anti-Islamic hackers (ethnoreligious). The most interesting part was the effect of the internet on war coverage. The war itself was dubbed as the first internet war, and the use of 'blogs' helped ordinary people to become involved in reporting the war and presenting an alternative by means of independent media, avoiding the restraints the corporate media faced, namely censorship and the demands and politics of advertisers (Media Advisory 3 April 2003).

The internet's role in the organization of anti-war protests

The importance of the internet in the organization of political groups is not news. However, in this particular conflict, its effects were for the first time indicative of the full potential of the new medium in politics. The months preceding the actual war in Iraq witnessed a plenitude of phenomena on, off and because of the internet that in previous international conflicts were only embryonic. Anti-war groups used email lists and websites, group text messages and chat rooms to organize protests, making politics more accessible to an unprecedented number of people from all backgrounds, who normally would not or could not get involved to such a degree.

In fact, anti-war protests in world capitals were impressive, whenever they actually made it to the newsrooms. In world capitals, people of all ages and nationalities took to the streets to demonstrate against the possibility of a war with Iraq. In only one weekend of 15 and 16 February 2003 (or F-15, in activist parlance), about ten million people protested globally against the war, rendering them the biggest peace protests since the anti-Vietnam War protests of the 1960s/70s (Gumbel 16 March 2003). In the US, the two biggest demonstrations took place in San Francisco and Washington. The disparity of protestors, where London is a case in point, is a sign that the anti-war movement has gone mainstream, thanks to hundreds of anti-war websites and mailing lists. The internet speeds up organizing, doing in months what took years in the Vietnam era (Max, Democrat and Chronicle). As historian and columnist Ruth Rosen explains in the *San Francisco Chronicle*: '[n]ever before in human history has an anti-war movement grown so fast and spread so quickly. It is even more remarkable because the war has yet to begin. Publicized throughout cyberspace, the anti-war movement has left behind its sectarian roots and entered mainstream culture' (Kahney 21 January 2003).

More astonishingly, millions of people all over the world were protesting the possibility of a war that had not yet started. According to Sarah Sloan, an organizer with International ANSWER, the internet played a very significant role, because 'it made a major difference in getting our message out there especially because the mainstream media is not covering the anti-war movement'. The internet also allowed protests to go international, with protestors in 32 countries holding demonstrations. 'There is no way the event would have been international without the Internet' (Kahney 21 January 2003). The same is being suggested by Rayman Elamine, organizer with Direct Action to Stop the War, an umbrella organization for a number of anti-war groups based in the San Francisco Bay area. 'Groups wouldn't have been able to do some of the logistical and other planning without the aid of the Internet for getting the message out' (Glasner 19 March 2003). Alistair Alexander from the Stop the War Coalition in England, has commented, in a *Guardian* article, that the web 'has allowed Stop the War to connect with people in a way politicians have failed to do. The much hyped age of online politics has finally arrived' (Alexander 19 February 2003).

Indeed, there is no end to practitioners and theorists discussing the possibilities of the medium in the organization of the anti-war movement. Kahney, for instance, mentions the United for Peace website, which includes news, contacts of activist groups and travel arrangements to the protests from 3,000 different cities. 'Before the Internet people felt blacked out by the media, because it doesn't represent their views. Now because of the Net, they feel like they are part of a movement. They are no longer isolated. It helps mobilize people, gets them to move' (Kahney 21 January 2003). Howard Rheingold, a well-known researcher in the field, has similar views. 'Instead of having some hierarchical top-down coalition, it's possible to have loose coalitions of small groups that organize very quickly' (Glasner 'Protests to Start When War Does').

In fact, the distinguishing characteristic of these groups is their disparity and their full use of networking, where we witness 'a mass mobilization without leaders – a digital swarm' (Bennett 17 March 2003). In other words, it exhibits the characteristics of a rhizome (Deleuze and Guattari 1987):

> What they (virtual networks) fostered was a form of interaction that preserved the integrity and autonomy of the constituent parts. No group was subject to the will of another. No group had to recognize one as a leading group or as the 'vanguard' of the movement. There was no need for bureaucracy, permanent staffs, officials, 'leadership', or even premises, beyond somewhere to house a server. Here was a form of interaction that denied the need for the very institutional and logistical framework that had, for a century defined the terms and conditions of political activism.
>
> (Tormey 2004: 65)

Reports of these protests tend to confirm this impression. Protesters were very graphically reported as

> [n]ot just the usual left-liberal suspects with their tie-dyes and political correct slogans, but Spanish-speaking bus drivers, public health workers,

suburban mothers and their children, blue-collar production line workers, lawyers and Republican-voting executives. Also unprecedented is the participation of the big labour unions, who where notoriously quiet during the Vietnam war . . .

(Gumbel 16 March 2003)

Thus the internet has become more than just an organizing tool. It can be argued that, by allowing mobilization to emerge from free-willing amorphous groups, rather than top-down hierarchies, the net has changed protests in a more fundamental way. 'It took four and a half years to multiply the size of the Vietnam protests twenty fold. This time the same thing has happened in six months', Todd Gitlin, sociology professor at Columbia, has commented (Lee 23 February 2003).

Virtually, the evidence of this is seen in the growing number of web pages supporting or criticizing the war. For instance, Moveon.org, a political website with 650,000 subscribers based in Silicon Valley, raised $400,000 through 10,000 or more individual donations to remake the 1960s 'Daisy' anti-nuclear war ad (Kahney 21 January 2003). Also, online groups promote and offer updates on protests, list event information, use chat rooms, conferences and email lists, and offer special updates on the conflict (www.notinourname.net, www.stopwar.com etc.). These include religious sites, with examples ranging from the Catholic church (www.vatican.va) and the US Conference of Catholic Bishops (www.usccb.org) to Jewish and Muslim sites, such as the aforementioned IslamiCity[1] (Webb Washington Post). The National Council of Churches provides a grass-roots toolkit on its site, which puts it this way: '[h]ere is a grassroots tool, roughly based on the Sherrod Brown Amendment, that activists can use to request information from their member of Congress on Iraq' (Webb Washington Post).

In Europe, anti-war groups are using the internet to organize protests outside US military bases and to organize protests generally, in countries such as Germany (www.resistthewar.de), Britain (www.reclaimthebases.org.uk, www.peaceuk.net), France (www.mvtpaix.org) and Spain (www.pazahora.org). Internationally, there are sites originating from Australia, South Africa, Egypt and also the global Independent Media Center (www.indymedia.org), which is linked to worldwide anti-war coverage, or sites like 'The Campaign Against Sanctions in Iraq' (www.casi.org.uk) (Webb 17 January 2003).

Not surprisingly, those supporting the war also rallied around the flag online. Organizations supporting the war used message boards, weblogs, online petitions and email to rally support for using force to disarm Iraq, from sites such as grassfire.net (Webb 14 March 2003).

More interestingly, there was a 'Virtual March' on Washington where tens of thousands of people bombarded the switchboard of the White House and other US government offices with protest calls and emails, halting much business in the capital. The coalition of 32 organizations which organized the action claimed that more than 400,000 people registered to participate in the campaign.

Meanwhile, 700 theatre groups in 42 countries led coordinated readings of the Aristophanes' anti-war comedy Lysistrata (Kahney 21 January 2003).

However, optimism is not ubiquitous. Stewart Nusbaumer, coordinator of Veterans Against the War, makes a somewhat valid point: 'On the one hand, [the internet] gives you larger numbers of people. But I've also noticed it's not great for a specific demonstration somewhere. I get emails from people who say: I live 2,000 miles away' (Glasner 'Protests to start when war does'). Still, there is recognition that the internet has been important in the growth of the movement. 'It has been invaluable in sharing information. With a growing constituency in the US becoming convinced the American Media is not giving an honest or complete picture, people have turned to alternative sources online, notably the European press' (Gumbel 16 March 2003).

The internet strengthens these types of movements by preserving the particularity of distinct groups and causes, while greatly facilitating the creation of networks of the like-minded:

> As well as preserving a distinct space or presence, groups could make common cause with other groups that shared their values. This could be achieved either through a simple 'links' page which indicated which groups they felt some sense of common cause with or through more elaborate networks, sites and mechanisms that acted as an umbrella organization all of its own.
>
> (Tormey 2004)

On the other hand, there are weaknesses in this kind of politics, since it could be argued that the leaderless and dispersed nature of online activism is ineffective, in that it ultimately fails to reach the vast majority of the world, where many activists in developing countries have little or no access to the internet or ICTs in general.

The internet's effect on media coverage

The internet was influential in the media coverage of this war for a variety of reasons, some touching on the simple fact that more people are online now than in the 1991 Iraq war, others emerging because technology has advanced to include instant messaging, audiovisual imagery, file sharing, etc. Still, what made the difference in this conflict was that the media environment ripe for alternative reporting of the conflict, mainly because Americans (as well as the rest of the world) searched online for news they could not find at home. An indicative example is that, as early as January 2003, according to *Wired News*, half of the 1.3 million visitors to the websites for Britain's *Guardian* and *Observer* newspapers were from the Americas (Kurtz 22 March 2003).

This part of the discussion looks at three main aspects of the internet's impact on war coverage: the integration of the internet into mainstream media, the effect of online material challenging official government sources and mainstream media, and the 'blogging' phenomenon, whereby everybody can be a journalist on the net.

'You are combining the speed of television with the depth of print. This could define how future wars are covered', Mitch Gelman, executive producer of CNN, the media network that defined the coverage of the 1991 Gulf war, commented (Swartz 18 March 2003). The web pages of all major media networks were well prepared for the integration of the internet in their coverage of the war. CNN itself planned for a war-tracker page, continuously featuring live reports from the frontline, 3-D charts that track bombs dropped, Iraqi casualties and defections and interactive maps of the battlefield, troop movements and terrain. ABCnews planned for fixed cameras on locations in Kuwait and Qatar 24 hours a day. MSNBC.com said that chat rooms and satellite transmitters that run off car cigarette lighters would be used, CBSNews.com would offer free video, maps and backgrounds and reporters with online notebooks and video feeds, and FoxNews would include the War on Terror page with a correspondent tracker and video clips from the front (Swartz 18 March 2003). As for the BBC's online coverage, it competed remarkably well with its American counterparts. On the downside, many independent news operations on the internet have slimmed down or disappeared altogether, since the info-bubble burst in 1999 (Reuters 6 March 2003).

Despite all the available tools, a large number of media watchdogs, journalists and audiences have protested the unchallenging position of the mainstream media towards both the decision to go to war, and in terms of the actual coverage during the war. As a result, the internet itself was used not only to mobilize international civil society as explained above, but also to offer alternative coverage of the conflict. There are different lines of development here: the US government's troublesome if not 'bombastic' relationship with the media, American media mostly following the government line with patriotic fervour, Americans turning to non-US sources by using the internet, and the rest of the world discovering the unpredictable and amazing effect of the internet on coverage, and the potential for first-hand eyewitness accounts via emails and blogging.

As far as the US administration is concerned, even though it made it clear there would be no censorship, it was very difficult for war correspondents that were not embedded with their troops to get non-official stories out. On the other hand, embedded journalists were controlled by the military. 'We will tell you what you can report from the speech afterwards', an army media organizer told journalists on their first day as embedded correspondents with 1st Fusiliers Battle Group (Tomlin 16 May 2003).

Embedded journalists in Iraq topped 800 at the height of the combat in 2003, but their number has since dropped to double figures. Five journalists have been kicked out of embedded slots for reporting secure information (Strupp 7 January 2005). Most US journalists do not leave their hotels and, in some cases, even their rooms are located in heavily fortified compounds in and around the Green Zone, the US military's Baghdad enclave. Their reporting is based in large parts on handouts from the US occupation officials, or material gained while 'embedded' with US military units, and is supplemented by on-the-spot accounts and interviews obtained by Iraqi 'stringers', who risk their lives for a fraction of the salary paid to their Western counterparts (Strupp 7 January 2005). Colonel Steve Boylan, a spokesman for the US military forces, acknowledged that some of the detained journalists have been 'held for several months'. None of them have been formally charged with any crime or even presented in court (Van Auken 7 May 2005).

This 'difficulty' demonstrates itself in the fact that the largest single group of war correspondents appears to have been killed by the US military. As Philip Knightley, writer of *The First Casualty: The War Correspondent as Hero and Myth Maker from Crimea to Kosovo*, puts it: 'the figures in Iraq tell a terrible story. Fifteen media people dead, with two missing, presumed dead. If you consider how short the campaign was, Iraq will be notorious as the most dangerous war for journalists ever' (Knightley 14 June 2003).

In a single day on 8 April 2003, a US missile hit an al-Jazeera office, killing a Jordanian journalist, and a US tank fired a shell at the Palestine Hotel, killing two more. Al-Jazeera offices in Basra were shelled on 2 April and a car clearly marked as belonging to the same station was shot at by US soldiers a day before the Palestine Hotel incident. International journalists and press freedom groups have condemned the attacks on the press corps in Iraq. 'We can only conclude that the US Army deliberately and without warning targeted journalists' (Reporters without Borders). 'We believe these attacks violate the Geneva conventions' (Committee to protect journalists in a letter to Defense Secretary Donald Rumsfeld 8 September 2003). The attacks on journalists 'look very much like murder', Robert Fisk of the London *Independent* reported on 3 March 2003 (www.fair.org 10 April 2003).

Eason Jordan, CNN's chief news executive, suggested at the World Economic Forum in Switzerland that some of the 63 journalists killed in Iraq had been specifically targeted by US troops (World Net 20 May 2005). Jordan quickly backed off his suggestion, but constant exposure from political weblogs led to his resignation. A year before that, he had admitted that CNN withheld news of atrocities taking place in Iraq under the regime of Saddam Hussein because the network was afraid it might lose access to the country. Echoing the same claim, the president of the 35,000-member Newspaper Guild asserted US troops are deliberately killing journalists in Iraq (World Net 19 May 2005).

Furthermore, with 600 correspondents, including about 150 from foreign media, accepting the Pentagon offer to be embedded with military troops, one would expect satisfying coverage. But even when embedded reports not consistent with the official Pentagon line appeared, they were not taken up from American media (Fair.org 4 April 2003). A look at some examples helps explain why more and more people turned to the internet for information on the war. A *Washington Post* article on William Branigin's eyewitness account describing the killing of civilians outside the Iraqi town of Najaf, where military procedures may not have been properly followed, was not picked up from the *New York Times*. Instead, *The Times* ran a story presenting the official line: 'Failing to Heed Warning, 7 Iraqi Women and Children Die' (Fair.org 4 April 2003).

Such uncritical coverage is hardly surprising, since several national and local media figures in the US found their jobs jeopardized, either explicitly or implicitly, because of the critical views they expressed on the war. Veteran war correspondent Peter Arnett was fired by NBC after giving an interview to Iraqi TV, Henry Norr was suspended without pay from the *San Francisco Chronicle* for using his sick day to get arrested in an anti-war protest, and Phil Donahue's talk-show was cancelled. MSNBC argued in an internal memo leaked to the All your TV website on 25 February 2003 that Donahue would be a 'difficult face for NBC in a time of war . . . He seems to delight in presenting guests who are anti-war, anti-Bush and skeptical of the administration's motives' (Fair.org 3 April 2003).

Again, this is to be expected in a media environment where official sources dominate US network newscasts and dissent is considered immoral.[2] Nearly two-thirds of all sources (64 per cent) used in news programmes were pro-war, while 71 per cent of US guests favoured the war. Anti-war voices made up only 10 per cent of all sources, just 6 per cent of non-Iraqi voices and a miserly 3 per cent of US sources. Viewers were more than six times as likely to see a pro-war source as one who was anti-war and, with US guests alone, the ratio increased to 25 to 1.[3] 'Given how timid most US news organizations have been in challenging the White House position in Iraq, I am not surprised if Americans are turning to foreign news services for a perspective on the conflict that goes beyond freedom fries', *Wired News* quotes former *Newsweek* contributing editor Deborah Branscum as saying.

> Although it's true that anyone with a website can publish news, it's still the established media players, such as newspaper publishers, that attract the largest share of an online audience, it's also true that more people are using the Internet as their primary news source, the same handful of companies run these sites.
>
> (Glasner 30 May 2003, quoting Murray)

The '100 Orders' penned by former US administrator in Iraq L. Paul Bremer include Order 65 passed on 20 March 2004 to establish an Iraqi communications and media commission. This commission has powers to control the media, because it has control over licensing and regulating telecommunications, broadcasting, information services and all other media establishments. The media commission sent out an order asking news organizations to 'stick to the government line on the US led offensive in Fallujah or face legal action' (Jamail 18 November 2004). It would be a worthwhile undertaking to look at http://www.iraqbodycount.net/resources/Fallujahh/index.php, which archives 300 selected news stories on the April 2004 siege of Fallujah.

In an article in the *Guardian*, Dahr Jamail, an unembedded journalist in Iraq, noted that refugees from Fallujah told him that 'civilians carrying white flags were gunned down by American soldiers. Corpses were tied to US tanks and paraded around like trophies' (Steel and Jamail 27 April 2005).

American documentary filmmaker Mark Manning returned from Fallujah after delivering supplies to refugees. Manning was able to secretly conduct 25 hours of videotaped interviews with dozens of Iraqi eyewitnesses. In an interview with a local newspaper in the US, Manning recounted how he was told grisly accounts of Iraqi mothers killed in front of their sons, brothers in front of their sisters, all at the hands of American soldiers. He also heard allegations of wholesale rape of civilians, by both American and Iraqi troops. Manning said he heard numerous reports of the second siege of Fallujah (November 2004) that described American forces deploying – in violation of international treaties – napalm, chemical weapons, phosphorous bombs and 'bunker-busting' shells laced with depleted uranium (Welsh 17 March 2005).

A *Los Angeles Times* scoop (3 June 2004) revealed that one of the most enduring images of the war – the toppling of the statue of Saddam Hussein in a Baghdad square on 9 April 2003 – was a US Army psychological warfare operation staged to look like a spontaneous Iraqi action: 'As the Iraqi regime was collapsing on April 9, 2003, Marines converged on Firdos Square in central Baghdad, as was widely assumed from the TV images – who decided to topple the statue, the Army report said. And it was a quick thinking Army psychological operations team that made it appear to be a spontaneous Iraqi undertaking' (Fair.org 12 March 2004).

As Gardiner sums it up:

> Among the fabricated stories was the early surrender of the commander and the entire 51st Iraqi mechanized division. We were told of an uprising in Basra – it did not happen. We were told on White House and State Department web sites that Iraqi military has formed units of children to attack the coalition – untrue. We were told of a whole range of agreements between the French and Iraq before the war over weapons – false. We were told Saddam had marked a red line around Baghdad and that when we crossed it Iraq would use chemical weapons – completely fabricated.
>
> (22 September 2004)

Again, in 'psy-op' terms, on the evening of 14 October, a young Marine spokesman near Fallouja appeared on CNN and made a dramatic announcement. 'Troops crossed the line of departure', 1st Lt Lyle Gilbert declared, using a common military expression signalling the start of a major campaign. 'It's going to be a long night' (Mazzetti 24 January 2005). CNN, which had been alerted to expect a major news development, reported that the long-awaited offensive to retake the Iraqi city of Falloujah had begun. In fact, the Falloujah offensive would not kick off for another three weeks. As Mazetti explains, Gilbert's carefully worded announcement was an elaborate psychological operation – or 'psy-op' – intended to confuse insurgents in Falloujah and allow US commanders to see how guerrillas would react if they believed US troops were entering the city, according to several Pentagon officials (Mazzetti 24 January 2005).

A report by the Defense Science Board, a panel of outside experts that advises Defense Secretary Donald Rumsfeld, concluded that a 'crisis' in US 'strategic communications' had undermined American efforts to fight Islamic extremism worldwide. The study cited polling in the Arab world that revealed widespread hatred of the United States throughout the Middle East. A poll taken in June by Zogby International (www.zogby.com) revealed that 94 per cent of Saudi Arabians had an 'infavorable' view of the Unite States, compared with 87 per cent in April 2002. In Egypt, the second largest recipient of US aid, 98 per cent of respondents held an unfavourable view of the United States (Mazzetti 24 January 2005).

In May 2004, photos of Iraqi prisoners being humiliated by US soldiers popped up all over the web as the internet once again proved to be the place millions of people turned to get information on a big story. While American newspapers were careful about how many and which prisoner photos they printed, lots of websites posted as many images as they could find in great graphic detail. Among the sites were www.thememoryhole.org and www.glob-alsecurity.org, a global think tank, which posted the full text of the army's report into the Abu Ghraib abuse and other documents on the Iraq prison scandal. Among the many activist sites covering the prison scandal were ElectronicIraq (electroniciraq.net) and AlterNet.org. Weblogs also posted exhaustive commentary on the naked pictures that were first publicized on CBS's '60 Minutes II'. English-language website al-Jazeera (english.aljazeera.net) published more subdued coverage, such as a photo gallery showing Muslims protesting outside the prison, and a survey of site visitors in which 62 per cent of the 72,840 respondents said they suspected that abuse of Iraqi prisoners was routine (Walker 9 May 2004).

Paul Taylor explores the extent to which the 'mental atmosphere' of the Abu Ghraib prison reflects more widespread values within the western mediascape that have become increasingly synonymous with the idea of public discourse (Taylor web paper). As Taylor very intelligently puts it, Baudrillard's notion of the ecstasy of communication was implicitly acknowledged by Donald Rumsfeld, who complained that it was much harder nowadays to control information sent back home by soldiers serving overseas. Unlike conventional letters, in which the censors can black out the offending parts, Rumsfeld bemoans the fact that US soldiers were 'running around with digital cameras and taking these unbelievable photographs and then passing them off, against the law, to the media, to our surprise' (Taylor quotes Sontag 13 May 2004).

Specifically in the light of Abu Ghraib, Sontag points out that although 'trophy' pictures have been taken in many previous military and social conflicts, these particular photographs:

> ... reflect a shift in the use of pictures – less objects to be saved than evanescent messages to be disseminated, circulated ... now the soldiers themselves are all photographers – recording their war, their fun, their observations of what they find picturesque, their atrocities – and swapping images among themselves, and emailing them around the globe, it was all

fun. And this idea of fun is, alas, more and more – contrary to what Mr. Bush is telling the world – part of the 'true nature and heart of America'.

(Sontag 13 May 2004 as quoted by Taylor)

It is worth mentioning how online newspapers operate here. With the net able to supply information almost instantly, newspapers have little choice but to put everything they publish on the internet daily, in the hope of keeping people on their site and returning to their site the next day. The struggle has always involved how to make money by charging for the content, while also keeping as many people as possible visiting the site to make it attractive to advertisers, with the ever-present back-of-the-mind fear that free and diverse online content will stop people buying printed newspapers.

According to K. McCarthy, newspaper websites and their content have gradually split into six areas. First, news stories which are free and will always be. After a week, though, these stories become archive stories and access to them may be billed. Second, columnists and opinion pieces news items that are exclusive and identifiable to the individual paper. Third, email services – sending directly to users' inboxes a concise rundown of stories that are likely to interest them. Fourth is the digital facsimile of the printed newspaper – whether in internet-standard jpeg images, PDF files or using some proprietary software. And finally, there are the add-ons: crosswords, competitions, games, etc. (McCarthy 9 July 2003).

What is more interesting, though, is the difficulties big media corporations have adapting to the digital age. The problem with creating new business models for the internet is that these can cannibalize existing, more lucrative businesses. The internet is emerging as a distribution system in its own right. Newspapers are suffering from the internet more than the rest of 'old media', as classified advertising moves online and young people use the net to get their news. Rupert Murdoch's News Corporation surprised everyone by buying Intermix Media, owner of MySpace.com, a social-networking site, for $580m, then Scout.com, a college sports site and IGN Entertainmant, a video gaming and entertainment site for $650m (*The Economist* 21 January 2006).

To continue, MoveOn.org, already mentioned for its successful anti-war efforts, concluded the same in a recent advertising campaign, which featured media mogul Rupert Murdoch under the banner, 'this man wants to control the news in America', claiming that Murdoch's News Corp, Disney, Viacom, GE and AOL Time Warner control 75 per cent of the total television audience and 90 per cent of the television news audience for broadcast and cable in the United States (Glasner 30 May 2003). In light of this, it is worrying that the Federal Communications Committee is considering scrapping decades-old regulations that have kept one or two companies from dominating the news.

In the first week of the war, internet traffic ran at twice the usual rate, according to ComScore Media Metrix. The at-work audience reached 36.5 million people on that particular Wednesday, almost matching the home audience of 37.1 million. Yahoo.com was among the sites whose usage skyrocketed. The volume of traffic to its news selection jumped 600 per cent on Thursday and Friday, and CNN.com had the most traffic of all news sites: nine million visitors. MSNBC was next with 6.8 million. The website of Britain's BBC drew nearly half a million visitors from the US alone on the Sunday of the same week, 60 per cent more than usual. The BBC site drew a worldwide audience of 3.1 million visitors on Sunday, while the top news site, CNN, drew 4.3 million (Walker 26 March 2003). Traffic also increased rapidly on anti-war websites, where on average, three leading protest sites (www.antiwar.com, www.united-forpeace.org and www.stopwar.org.uk) drew 160 per cent more traffic than they had four weeks previously (Walker 21 March 2003). Also, according to Hitwise's media alerts service, which trawls 11,000 articles daily (covering the top 300 global news and media sites), 40 per cent of articles that week related to the war against Iraq. For the week ending 22 March, news websites accounted for a 22.47 per cent share of all web traffic in the UK, an increase of 9 per cent (Europemedia.net 'War dominates the Web').

Moreover, three-quarters of online Americans (77 per cent) have used the internet in connection with the war in Iraq. More than half of the nation's 116 million adult internet users have used email to communicate or learn about the war (Rainie *et al.* undated). According to the PewInternet study, 17 per cent of online Americans say their principal source of news is the internet and in the days before the war broke out, 37 per cent of internet users got news on a typical day. Interestingly, war opponents are slightly more likely than supporters to report intensified internet use.

Another contribution of the internet to this war was in the form of 'blogs'. Blogging is an easy and fast way for personal publishing on the web. Some examples are indicative of the potential for citizen journalism. During extensive flooding, CNN, the BBC and others received powerful firsthand accounts by email long before camera crews and correspondents were on the scene. The email newsgathering was vivid and also included colour stills, which were posted into website galleries hours before newspapers published their accounts[4] (Cramer 9 March 2003). To a lesser extent, blogging was experienced during the NATO bombing of Belgrade in 1999.

In South Korea, OhmyNews, which publishes 200 stories a day, mostly written by more than 26,000 registered citizen journalists, has two million daily readers and has been widely credited as helping elect South Korea's new Prime Minister (Kahney 17 March 2003). By some measures, South Korea is the most wired country in the world, with broadband connections in nearly 70 per cent of households. Around election time OhmyNews was registering 20 million page views per day. The service averages about 14 million visits daily, in a country of only about 40 million people.

In June 2003, Iranian blogs gave voice to dissidents (www.hoder.com, www.iranian.com) while a growing network of Iranian-American media outlets have been aiding the student-led protests. In another instance, bloggers reported alternative news from a G8 meeting[5] (Theodoulou 26 March 2003). During the Iraq conflict, a blogger called Salam Pax,[6] blogging from Baghdad

(Dear_raed.blogspot.com), and Christopher Allbriton, a New York-based veteran journalist (Back-to-iraq.com), have been profiled in many news stories. Reporters of *Time* magazine, the BBC and other leading news outlets had their own blogs. Kevin Sites, a CNN correspondent, posted pictures, audio and commentary on his website from the Kurdish section of Iraq. CNN asked Sites to suspend the blog.

The effect blogging had on coverage might not have been profound in this war,[7] but it was an indication, perhaps, of where war coverage might be going.

> For all the saturation coverage of the invasion of Iraq, this has become the first true Internet war, with journalists, analysts, soldiers, a British law-maker, an Iraqi exile and a Baghdad resident using the medium's lightning speed to cut the fog of war. The result is idiosyncratic, passionate and often profane, with the sort of intimacy and attitude that are all but impossible in newspapers and on television.
>
> (Kurtz 22 March 2003)

Kurtz also cites law professor Glenn Reynolds, whose site, InstaPundit.com, saw a surge in traffic as the Iraq crisis has heated up, doubling to 200,000 hits a day:

> The most interesting thing about the blog coverage is how far ahead it is of the mainstream media. The first hand stuff is great. It's unfiltered and unspun. That doesn't mean it's unbiased. But people feel like they know where the bias is coming from. You don't have to spend a lot of time trying to find a hidden agenda.
>
> (Kurtz 22 March 2003)

Because of their personal nature, weblogs have served as a great filter for the pro- and anti-war lobbies, but apart from the now globally famous Iraqi blogger Salam Pax, very few of these sites actually gives new information. The wire services, broadcasts and newspapers have the most journalists close to the fighting and are able to fire reports off instantly through TV, radio and the web. And this is, perhaps, where electronic media have made the biggest difference. The BBC's very successful rolling weblog of all correspondents in the Gulf and the US carried a mention of the missile hitting the Baghdad marketplace before it made it into the air (Bell 30 March 2003).

However, the global internet audience is still in its earliest stages. Only about 5 per cent of the world's population can access the internet, according to Nielsen/Net Ratings. It is hard to fathom what will happen to world opinion should net use reach 50 per cent or more (Walker 21 March 2003). Notably, UN Secretary-General Kofi Annan, before the World Electronic Forum, a summit of world leaders in December 2003, said that the focus should be on expanding internet use and reaffirming media freedoms and the rights of ordinary people to stay informed.

The information summit centred on whether the United Nations should have more control of the internet, since the key decisions are made by a private, US-based organization of technical and business experts known as the Internet Corporation for Assigned Names and Numbers (ICANN), and as well as who will pay for getting more poor nations online. The broadcasters asserted that the future is not only online, since radio and TV will remain the dominant means of mass communication in many poor countries for decades (*Wired* 9 December 2003). One of the main areas of conflict centred on who should pay for technology projects in the developing world. African nations have been rallying behind a proposal from Senegal to set up a new 'digital solidarity fund'. Many industrialized nations are wary of creating a new UN fund, instead supporting investment by private companies and redirecting existing funds (BBC Online 29 September 2003).

The impact of the war on the internet itself

The possibility and actuality of war with Iraq instigated cyberattacks between pro-islamic/anti-islamic hackers (ethnoreligious cyberconflict) and pro-war/anti-war hackers (sociopolitical cyberconflict). Despite the official US government warning against patriotic hacking, the most notorious incident occurred when the al-Jazeera website was knocked offline by an American web designer (Associated Press 12 February 2003). In an incident originally attributed to the website's inability to deal with traffic, the site was disabled by hackers for long periods of time (Knight, Newscientist.com). The Arabic news broadcaster's domain name was redirected to patriotic web pages or porn sites.[8] According to Ballout, an employee of Network Solutions was tricked into giving the culprit a confidential password that allowed the hacker to temporarily assume total control of al-Jazeera's domain (Delio 21 March 2003). Al-Jazeera became the internet's number one search query for 48 hours, according to web portal Lycos. The reason was that people were hunting for video footage that al-Jazeera had aired of dead American soldiers, and U.S. prisoners being interrogated by their Iraqi captors – including gruesome images that American TV networks mostly declined to show. Most foreign news sources[9] noted that the US media had shown images of Iraqi prisoners before demurring and showing the American prisoners of war (Walker 26 March 2003).

Yet this only gives a small taste of what really happened before and during the war. Cyberattacks were occurring as early as October 2002, debating virtually the situation in Iraq. London-based computer security firm mi2g said October 2002 was the worst month for digital attacks since its records began in 1995. It estimated 16,559 attacks were carried out on computer systems and websites during that month. But the computer security firm said the economic damage caused by the attacks is decreasing, reflecting a decline in the quality of targets chosen. According to mi2g, which monitors the hacking of websites, the number of attacks by groups opposed to action in Iraq, as well as Israeli attacks on Palestinians, rose tenfold that month. 'We have noticed that more and more Islamic interest hacking groups are beginning to rally under a common anti-US, UK, Australia, anti-India and anti-Israeli agenda. The most active hacking groups are USG, with members from Egypt, Morocco and Eastern Europe, and FBH which is based in Pakistan' (BBC News 29 October 2003). Unix Security Guards (USG) defaced nearly 400 websites in a single day with anti-war slogans written in Arabic and English, according to iDefense (Krebs 20 March 2003).

When the war actually started, Zone-H, a firm that records and monitors hackings reported 20,000 defacements in the first week of the war. Hundreds of US and British business, government and municipal websites were defaced with anti-war messages, security experts reported. Seemingly within hours, more hawkish hackers went on the offensive against Arab sites. Roberto Preatoni, founder of Zone-H, commented at the time: 'this is the future of protest. If you take down al-Jazeera, everybody around the world knows it. And you never have to leave your house' (Reuters 28 March 2003).

As a result of the escalating conflict, thousands of websites were the target of Denial-of-Service attacks, defacement, worms and viruses[10] (Delio 21 March 2003). According to F-Secure, another security firm, the majority of deface-ments were from anti-war hackers, with anti-war messages or images, including those comparing the physical similarity between US president George Bush and a monkey, scrawled across various homepages. On the Sunday after the initial bombing, 10 Downing Street's website was inaccessible, having been the target of a distributed Denial-of-Service assault (Europemedia 1 April 2003). Then, in mid-June, hackers put up a picture of President Bush carrying his dog, with Prime Minister Tony Blair's head superimposed on it. (Reuters 16 June 2003) 'At the moment we are tracking over a thousand such defacements, most with anti-war messages', commented Jason Halloway of F-Secure. 'I have never seen that level of political hacktivism before, nor so many defacements in such a short time' (Reuters 16 June 2003).

Defacements and denial of service were accompanied by at least three war-related worms – Ganda, Lisa and Wanor – which shut down security, deleted critical system files or erased hard drive data (Krebs 20 March 2003). However, according to Symantec, most attacks, including one in October 2002 that brought down nine of the 13 servers that support the internet, cannot get around the fact that, when online traffic is disrupted or blocked in one place, it tends to flow through thousands of alternative channels instead (McMillan 27 March 2003). With most netwarriors wanting the internet up and running, such attacks are symbolic and do not aim to bring down the net. Symantec itself has been at the centre of controversy numerous times and in different cases – as mentioned throughout this book. In this particular anti-war cyberconflict, American online activist David Swanson says ISP Comcast, and security services company Symantec, blocked emails with 'www.afterdowningstreet.org' in the body of the email for a week. The emails were drawing attention to the so-called Downing Street memo, first published in *The Times* newspaper, which shows that the Iraq war was planned well in advance. A spokeswoman for Symantec said that a spam rule was created due to an increase in email traffic, but was later turned off for being too broad (Varghese 27 July 2005).

Nevertheless, the truth of the story is that such activities are not appreciated by hackers, who in many instances have served as scapegoats for script kiddies[11] with moderate computer skills. Oxblood Ruffin, director of Hactivismo, a group that develops tools to circumvent censorship, has commented that 'the individual(s) who did this are committing a computer crime and causing censorship' (Delio 31 March 2003). Robert Ferell, a security researcher, adds that '[m]ost of them have no clear grasp of the causes they are supposed to be supporting or fighting against. They just want to appear "hacktivists", because that's a cool label to have' (Delio 31 March 2003). And Mark Loveless, a hacker working for US security software company Bindview, put it this way: '[i]n a protest or activist scenario, one would hope that one's cause and message were strong enough that "shouting down" the opposing viewpoint is considered unnecessary' (Reuters 28 March 2003).

In fact, the US government used the internet for the first time in a campaign aimed at Iraqi email addresses, spamming recipients to contact the UN if they wanted to defect. Saddam responded by shutting down internet service providers. (Cramer 9 March 2003). The US action was quite a paradoxical venture, since only 12,000 of Iraq's 12 million people were online. Also, the US was censoring troops' emails to family and friends, to prevent leaks of sensitive information. Yet the real potential of cyberwar for the US military probably lies in the military use of a highly secure intranet and wireless systems that speed audio, video and other data back to command and control systems, as noted by Winn Schwartau, a cyberwar expert (Tsuroka 17 March 2003).

US diplomacy officially entered the electronic age with the completion of a two-year project to provide internet access to all US embassies and consulates, with some 44,000 foreign service officers and other embassy staffers able to access the web from Washington's more than 260 far-flung diplomatic missions, ranging from Afghanistan to Zimbabwe (Yahoonews 29 October 2003). Accordingly, the US Department of State is well aware of the information warfare threat, as are the diplomatic departments of the top industrial nations. However, the extent of knowledge of the average diplomat, embassy and consulate staff around the world remains questionable.

It was also the first war 'in which thousands of hours of digital imagery will be shot of actual combat.[12] A lot will be from unmanned aerial vehicles flying over the battlefield' (Lee 17 March 2003). As Der Derian notes, the Pentagon's current gospel is network-centric warfare, an observation also found extensively in the work of Arquilla and Ronfeldt (McClellan 19 February 2003). As Der Derian graphically puts it:

> Command and control networks, like the air defense networks, will be taken out with missiles and possibly even electromagnetic pulse weapons. Prime time/cable networks will be red hot with war fever and coverage. NGOS will roll out humanitarian networks. Anti-war networks will send out marching orders. Soldiers, sailors and airman will email stories back home. This war will be started and ended by networks.
>
> (19 February 2003)

Conclusion

Following the integrated theoretical model proposed in pages 86–93, relevant links can be made with the empirical analysis undertaken above. Interestingly, all the components of the framework are represented in this particular conflict. The first component on the environment of real conflict was not fully explored here, because it is explained more thoroughly in the theoretical sections. The virtual conflict environment was not fully explored either, because this is explained in general terms in Chapter 3, and in the Iraq war example in the section on pages 187–189 in this chapter.

Upon examination mobilization structures appear to have been greatly affected by the internet. Peace groups organized demonstrations and events through the internet, to the effect that ten million people protested against the war globally, with the net speeding up mobilization remarkably. It helped mobilization in loose coalitions of small groups that organized very quickly, at the same time preserving the particularity of distinct groups in network forms of organization.

Moreover, the framing process was affected as well, since email lists and websites were used to mobilize, changing the framing of the message to suit the new medium. The language used to mobilize through the internet differs from traditional political discourse (for instance, speeches or texts in traditional media) in that it can combine various technical media (video, satellite images, file-sharing) in a way that delivers on the one hand a richer message, but on the downside a sometimes hasty and crude, under-analytical political message. The political opportunity structure in this particular case can refer to the rise of alternative media (as we see below), but also to an opening of political space, and an opening of global politics to people who would not or could not get so involved before. In virtual terms, hacktivism was apparent in anti-war/pro-war hacking, anti-Islamic/pro-Islamic hacking and a Virtual March on Washington, which impacted on the city's communication infrastructure.

On the hacking front, pro-Islamic/anti-Islamic hacking is an example of ethnoreligious cyberconflict. The link between ethnoreligious affiliation and discourses of exclusion/inclusion is evident, when considering the al-Jazeera hack from American hackers, and the movement of Islamic hackers united in a common anti-US, UK, Australia, anti-Indian and anti-Israeli agenda. Furthermore, the use of the internet as a propaganda and mobilizational tool is common to both sides (anti- and pro-war), through a considerable amount of websites advocating one view or another and mobilizing, counter-mobilizing and anti-mobilizing against each other.

On the media front, it is clear that political discourse is constructed in the American mainstream media to mobilize support for the war as analysed above, since, for example, more than two-thirds of all sources in news programmes were pro-war. Also very important is the issue of alternative sources and censorship. Because of the embedded system, journalists having their work jeopardized for not being 'patriotic' enough, and the American media generally following the government line, Americans and the rest of the world went online to find alternative news and first-hand eyewitness accounts via emails and blogging. The result was the integration of the internet into media coverage and the distribution of online material challenging official sources. The Wolfsfeld model is comfortably applied in consideration that the anti-war groups had the ability to initiate and control protest events and to mobilize supporters, but were not as successful in dominating political discourse. The media effects on policy were, above all else, technical. As a result, there was instant 24-hour access to the war, bringing with it the pressure this would inevitably put on any administration. However, no actual debate or impact on policy took place, since the American media failed to question any decisions being taken by their government.

In the final analysis, the internet played a distinctive role in the spread of the peace movement, on war coverage and on war-related cyberconflicts, in relation to which the full potential of the new medium in politics was shown. In the months preceding the actual war in Iraq, a plenitude of phenomena on and off the internet emerged, which in previous international conflicts were only embryonic. Anti-war groups used email lists and websites, group text messages and chat rooms to organize protests and, in some cases, to engage in symbolic hacking against the opposite viewpont. The integration of the internet into mainstream media, the effect of online material challenging official government sources and the mainstream media, and blogging are possible indications of where war coverage might be going when internet users exceed their present numbers.

The most interesting thing in a brief application of the CC theoretical framework is the level of censorship and the Wolfsfeld model. In terms of censorship, the latest literature supports the idea that journalists were not only censored and manipulated (CNN incident), but also targeted in this conflict – which brings up the issue of whether the US could control information. Apparently, through psyops, they could manipulate the conflict and control the media, especially the American mainstream media (almost always submissive to the patriotic/nationalistic discourse after 9/11). Their mentioned inability to control inconsistencies and fiascos from 24-hour internet coverage, blogs and US soldiers using the internet to send pictures (as in the Abu Ghraib prison incident), or to manipulate the American image in the Muslim world, is nevertheless another issue. Accordingly, the anti-war movement succeeded in that respect at gradually building their own image of the Americans and their allies and framing their message (no WMD, dodgy dossiers, humanitarian concerns, etc.).

7 Conclusion

This book has addressed the problem of political conflicts in computer-mediated environments (cyberconflicts). Firstly, the question was addressed of how modernist concepts like power, participation and democracy fit into a global postmodern medium, the internet, in an era of globalization. Following this were arguments and analyses of three theories: media theory, social movement theory and conflict theory, as a way to understand the two kinds of conflict identified (ethnoreligious and sociopolitical), thereby deriving and proposing an integrated analytical framework for cyberconflict. Thirdly, an analysis was undertaken of the cyberconflict environment, by way of investigating the historical background for the phenomenon of cyberconflict, information warfare, cyberterrorism and internet security analysis.

In the empirical chapters, these two types of cyberconflict were analysed, and examples given, including the Israeli–Palestinian, the Indian–Pakistani and the anti-war and anti-globalization cyberconflicts, and including a discussion on cyberdissidents' conflicts with governments. The conclusions in the empirical chapters presented a possible application of this work's theoretical framework for cyberconflict to the empirical evidence, for an easier understanding of cyberconflict and its consequences for the political world.

Information communication technologies act as a force multiplier, enhancing power and enabling social actors to punch above their weight and attain a reach and influence previously denied to them. The groups that use the internet are able to communicate messages to a wider audience than that reached by more traditional means of political communication. The promise of nearly unlimited information delivered to a user's monitor in mere seconds is a promise for a better democracy, since the internet can help to make citizens more active and more knowledgeable about their governments. However, this availability of unlimited information may inspire, and indeed has inspired, those who are already politically active, but it does not necessarily mean that the internet alone will increase the numbers of the attentive public.

As far as democracy is concerned, network technologies are increasingly used in public and political debates and communications, thus promoting dialogue between opposing parties – one of the elements of true democracy. New developments in governance such as deregulation, creating independent agencies, privatization, and governing at a distance are factors resulting from ICTs. New uses of technology such as 'blogging' are the harbinger of a new interactive culture that could potentially change how democracy works, turning voters into participants rather than passive consumers. At the same time, however, technologies intruding upon privacy, surveillance and censorship may impede democratic liberties, due to their catastrophic effect on free speech and freedom of expression.

On the question of participation and the internet, providing greater choice and opportunity only solves the technical problems of participation – for example, that related to reducing the cost of involvement. It does not get to the heart of what motivates citizens to move from a state of disengagement to one of salutary involvement in civic life. In other words, universal access is not sufficient for realising a democratic polity.

On the question of political groups pursuing power, the internet offers opportunities for breaking down political hierarchies, while subverting the national boundaries that have helped, in part, to control flows of information. There is also a growing control of cyberspace by elites who are defined by their technical expertise – that is, the ability to alter the 'thingness' of technology that constructs online life, while the gap between the digital haves and the digital have-nots continues to grow. Also, cyberspace appears as a place in which individuals can put aside many of the inequalities of offline life, simply because nobody knows if they are 'really' female, old or disabled. In a way, cyberspace can redefine problems such as the broadening and democratization of decision-making procedures by removing the constraint of physical presence.

Furthermore, globalization brings dramatic changes in the transactions and interactions taking place among states, firms and peoples in the world. It involves both an increase in cross-border transactions of goods and services and an increase in flows of images, ideas and people. In such a terrain, the internet seems to be forming a 'cybernation', where its initial subculture evolved a set of acceptable behaviours, a common history and a common identity of beliefs: free speech, protection of civil rights, privacy and freedom of expression, etc. With the spread of the internet, there is scope for a newly international localism, which is finding expression in virtual communities, with some people going so far as to suggest that a new global cyberstate is forming.

When considering the postmodern nature of the internet, it cannot be forgotten that the resistance of new media to modernity lies in their complication of subjecthood, their denaturalizing of the process of subject formation, and their questioning of the interiority of the subject and its coherence. The issue of identity is not merely of philosophical importance, but is also an issue of immense practical importance for the conduct of states, not to mention identifying the lines of flight from the status quo. Moreover, the shift from real to hyperreal occurs in transit from mere representation to simulation, a movement which already exists in the virtual world. Once the internet moves closer to total connectivity, this metaphorical cyberspace could become the hyperreal – more real than the place it once simulated. The fact that many people believe virtual communities to be real places in which they live real experiences brings this blurring of the real and the unreal closer to Baudrillard's postmodern moment of hyperreal than to representation.

Searching for a satisfactory description of empirical cases of cyberconflict led to the use of a classification between two types of cyberconflict: sociopolitical and ethnoreligious. In order to explain the empirical evidence of 'cyberconflict', the integration of elements of social movement, conflict and media theories into a single analytical framework for cyberconflict was proposed. Elements of social movement theory were identified, to discuss sociopolitical cyberconflicts, conflict theory to address ethnoreligious cyberconflicts and media theory as a component for both, deriving a single integrated analytical framework for understanding cyberconflict.

In the social movement theory chapter, elements of this theory were identified, which were relevant to new social movements using the internet to accomplish their goals of power, democracy and participation. More specifically, the classical resource mobilization model of mobilizing structures was utilized. This enabled the framing of processes and the political opportunity structure, and the analysis of how these are affected by new social movements' use of the internet.

New social movements are not new but, rather, part and parcel of the dominant modern culture, which makes it difficult to think of movements as flowing either from 'pre-modern' or 'postmodern' subcultures. However, the structure of NSMs – open, decentralized, nonhierarchical – makes them ideal for internetted communication. The movement is composed of adverse autonomous units that expend an important part of their resources on internal solidarity. A network of communication and exchange keeps the cells in contact with each other. Information and resources circulate in networks, and leadership is not concentrated but diffuse. NSMs advocate direct democracy, self-help groups and cooperative styles of social organization. The fewer and weaker the social ties to alternative networks, the greater the structural availability for movement participation. Sociopolitical movements, such as the political dissidents in China, can test the limits of a system, pushing the system beyond the range of variations that it can tolerate without altering its structure.

Conflict theory was used to analyse the present situation of conflict between sociopolitical groups, as well as ethnoreligious ones. Neo-liberal governments and institutions face a counter-hegemonic account of globalization, to which they have responded in a confused and often contradictory manner. One of the interesting sides to the argument is that the information revolution is altering the nature of conflict by strengthening network forms of organization over hierarchical forms. In contrast to the closure of space, the violence and identity divide found in ethnoreligious discourses, sociopolitical movements seem to rely more on networking and rhizomatic structures.

In media theory, the important questions revolved around how information is released and why, how much censorship is taking place and what alternative sources of information are available. These questions were posed in an analysis of the anti-Iraq war/pro-war cyberconflict. The way a war is communicated is as important as the conduct of the war itself. Among many examples, the Moscow LiveJournal incident shows that individuals and protagonists can now send stories more quickly than journalists, indicating that the media will have an independent capability to access future conflict arenas and to provide real-time visual and audio coverage of battlefield events. This has consequences for news management, even by very powerful states like the US.

The analyses of these theories led to an integrated theoretical framework with the following parameters to be looked at while analysing cyberconflicts:

1   *Environment of conflict and conflict mapping (real and virtual).* The world system generates an arborescent apparatus, which is haunted by lines of flight, emerging through underground networks connected horizontally and lacking a hierarchical centre (Deleuze and Guattari). The structure of the internet is ideal for network groups (since it is a global network with no central authority) and has offered another experience of governance (no governance), time and space (compression), ideology (freedom of information and access to it), identity (multiplicity) and, fundamentally, an opposition to surveillance and control, boundaries and apparatuses. However, in ethnoreligious cyberconflicts, where the groups' systems of belief and organization aspire to hierarchical apparatuses (nation, religion, identification with parties and leaders), this network form is not always evident. This is why there is a dual modality of cyberconflict: one rhizomatic and one hierarchical.

2   *Sociopolitical cyberconflicts.* The impact of ICTs on: a) mobilizing structures (network style of movements using the internet, participation, recruitment, tactics, goals); b) framing processes (issues, strategy, identity, the effect of the internet on these processes); c) political opportunity structure (the internet as a component of this structure); d) hacktivism.

3   *Ethnoreligious cyberconflicts.* a) ethnic/religious affiliation, chauvinism, national identity; b) discourses of inclusion and exclusion; c) information warfare, the use of the internet as a weapon, propaganda and mobilizational resource; d) conflict resolution, which depends on the legal and organizational framework, the number of parties and issues, the distribution of power, and the content of values and beliefs.

4   *The internet as a medium.* a) analysing discourses (representations of the world, constructions of social identities and social relations); b) control of information, level of censorship, alternative sources; c) political contest model among antagonists – the ability to initiate and control events, dominate political discourse, mobilize supporters (Wolfsfeld); d) media effects on policy (strategic, tactical, and representational).

The political environment of the internet is analysed not in terms of the internet as a mass medium in the traditional sense but, rather, as a significant new resource used by the opposing parties in a conflict. This work's approach involves analysing the use of the internet by the parties in a conflict (endogenously) and not just theorizing about how the media influence the political outcome of a conflict (exogenously). What was evident from the beginning was that major political and military conflicts are increasingly accompanied by a significant amount of online activity. Furthermore, cyberattacks are escalating in volume, sophistication and coordination.

The third chapter of this book illustrated the environment of cyberconflict placing emphasis on hackers, security experts and internet security analysis. Hacking can potentially perform a variety of benevolent services to the security industry, constantly pushing forward the limits of computer security, being an important form of watchdog counter-response to the use of surveillance technology and data gathering by the state. At the same time there is a knowledge gap between computer security and computer underground. There is a scarcity of theoretical knowledge surrounding computer security, with demand for more hands-on experience of security to supplement more formal theory.

A further part of the cyberconflict environment is cyberterrorism: computer-based attacks intended to intimidate or coerce governments or societies in pursuit of goals that are political, religious or ideological. According to Arquilla and Ronfeldt, conflicts increasingly revolve around knowledge and the use of soft power. This would come about with the help of information-age ideologies in which identities and loyalties shift from the nation-state to the transnational level of global civil society. Additionally, netwar is referred to as the low, societal type of struggle, while cyberwar refers more to the heavy information warfare type.

The introduction to the technical environment of cyberconflict was followed by description and analysis of specific cases of political conflicts online, which were then linked with the integrated analytical framework. Sociopolitical cyberconflicts were examined first. Sociopolitical cyberconflicts could be seen as taking two forms: first, when proper hackers attack virtually chosen political targets; and second, when people organize through the internet to protest or carry a political message through email. Sociopolitical CCs seem to rely heavily on decentralized networks and indicate the use of rhizomatic structure, following the structure of NSMs as discussed above.

Cyberconflicts can act as a 'barometer' of real life conflicts and can reveal the natures and the conflicts of the participating groups. The protagonists in sociopolitical cyberconflicts fight for participation, power and democracy. Evident in the anti-globalization and the anti-capitalist movement is an alternative programme for the reform of society, asking for democracy and more participation from the 'underdogs', be they in the West or in the developing world. In the anti-war movement, which is a single-issue movement, the demand is for a change in power relations in favour of those who believed the war to be unjustified. In new social movements, networking through the internet links diverse communities such as labour, feminist, ecological, peace and anti-capitalist groups, with the aim of challenging public opinion and battling for media access and coverage. Groups are being brought together like a parallelogram of forces, following a swarm logic, indicating a web of horizontal solidarities to which power might be devolved or even dissolved. The internet encourages a version of the commons that is ungoverned and ungovernable, either by corporate interests or by leaders and parties.

An early example of hacktivism is the Seattle anti-WTO mobilization at the end of November 1999, which was the first to take full advantage of the alternative network offered by the internet. The anti-WTO protesters were able to initiate a newsworthy event, putting their opponents on the defensive. Using the internet, they could send stories directly from the street for the whole world to see, rendering the flow of information uncontrollable. Thirdly, they were able to mobilize support by promoting an alternative frame for the event.

Dissidents against governments are able to use a variety of internet-based techniques (email lists, email spamming, BBS, peer-to-peer and e-magazines) to spread alternative frames for events and a possible alternative online democratic public sphere. An example of dissidents' use of the internet is spamming e-magazines to an unprecedented number of people within China, a method which provides recipients with 'plausible deniability'. Also, file-trading networks like Kazaa and Gnutella can help dissidents communicate, since they have no central source and are hard to turn off.

Ethnoreligious cyberconflicts primarily include hacking enemy sites and creating sites for propaganda and mobilizational purposes. Empirical evidence and analysis of such conflicts (Israeli–Palestinian, India–Pakistan, China–US, Taiwan–China–Japan, Kosovo, Colombia) were provided. In ethnoreligious CC, despite the fact that patriotic hackers can network, there is a greater reliance on traditional ideas, such as protecting the nation or fatherland and attacking for nationalist reasons. The Other is portrayed as the enemy, through very closed, old and primordialist ideas of belonging to an imagined community.

The brief section on conflict resolution on the internet looks at attempts and experiments and concludes that the internet is a helpful tool that fosters cooperation and dialogue between opposing sides in a conflict, bringing individuals and groups together that would otherwise have found it difficult to meet. The three examples examined in this work (MEVic, MEAF and the Project for Arab–Jewish Dialogue) place an emphasis on community-based diplomacy, interfaith dialogue, and building positive relationships as a platform for resolution.

The Israeli–Palestinian cyberconflict saw the use of national symbols (like the Israeli flag, Hebrew text and even a recording of the Israeli national anthem) when hacking the Hezbollah home page. This explicitly draws attention to issues of national identity, nationalism and ethnicity. Also, the language used by hackers relies on an 'us' and 'them' mentality, where Israelis and their American supporters, or else Palestinians and Muslims, are portrayed as barbaric, reflecting discourses of inclusion and exclusion. The internet in this cyberconflict became a battleground and was used as a weapon by both sides, and full-scale action by thousands of Israeli and Palestinian youngsters involved both racist emails and circulating of instructions on how to crush the enemy's websites.

Similarly, in the Indian–Pakistani cyberconflict, the Indian army's website was set up as a propaganda tool, and hacked pictures of alleged tortures of Kash-miris by Indians were placed on the site, in a similar propaganda tactic. Also, the internet was used as a weapon, when the worm Yaha was released by Indian hackers. In particular disourses, religion is mentioned (religious affiliation), the word 'brothers' (collective identity and solidarity) and 'our country', a promised land. In the rest of the examples, similar links can be made with the proposed framework.

In analysing the March 2003 Iraq conflict, the internet's role in the conflict was studied, in terms of its effect on the organization and spread of the move-ment, and its impact on war coverage and war-related cyberconflicts. These last involved hacking between anti-war and pro-war hacktivists (sociopolitical CC), but also between pro-Islamic and anti-Islamic hackers (ethnoreligious CC).

Mobilization structures, for instance, were greatly affected by the internet, since the peace groups used the internet to organize demonstrations and events, to mobilize in loose coalitions of small groups that organize very quickly, and to preserve the particularity of distinct groups in network forms of organization. Moreover, the framing process was also affected, since email lists and websites were used to mobilize, changing the framing of the message to suit the new medium. The political opportunity structure in this particular case refers to alternative media, but also to an opening of political space, or an opening of global politics to people who, previously, would not or could not get as involved.

This book identifies a duality of cyberconflict, where ethnoreligious cyber-conflicts are mapped as representing/defending loyalties of hierarchical appara-tuses and sociopolitical cyberconflicts are empowering network forms of organization. The thesis also promotes the argument that actors in ethnoreligious CC need to operate in a more *networked* fashion, if they are fighting network forms of terrorism or resistance. Actors in sociopolitical CC need to operate in a more *organized* fashion, if they are to constructively engage with the present global political system or parts of that system. Conflict resolution will be pos-sible only when hierarchical apparatuses become more networked, and when rhizomatic groups become more conscious of the rest of their hosting network. Ultimately, what this 'reversal' argument calls for is two-fold. First, sociopoliti-cal cyberconflicts need to see a strengthening of the organizational structures and informal weak connective structures which non-hierarchical movements normally have. As Tarrow explains, the most effective forms of organization are based on partly autonomous and contextually rooted local units linked by con-nective structures, and coordinated by formal organizations. Second, the information revolution is favouring and strengthening networked organizational designs, often at the expense of hierarchies. As Arquilla and Ronfeldt argue, and as 9/11 demonstrated, states need to wake up to this fact and realize that net-works can be fought effectively only by flexible network-style responses.

The impact of information technology on political conflict cannot be ignored. The form of the internet itself – a global network with no central authority – has offered another experience of governance (no governance), time and space (compression), ideology (freedom of information and access to it), identity (multiplicity) and fundamentally an opposition to surveillance and control, boundaries and apparatuses. New information-age ideologies could easily argue for a transfer of virtual social and political structures to the real world, reversing, for once, the existing process of imitating real life in cyberspace. The internet is not a medium. It is 'another' place. In this environment, cyberconflicts are complex conflicts involving social movement, conflict and media components, which in this book are combined to derive an integrated thereoretical framework to explain this novel and exciting phenomenon.

Glossary

Asterisks (*) indicate that the definition is taken from *The Online Hacker Jargon File 4.1.0*, the 27 October 2003 update.

Authentication   Verifying that a person is known to you.

Avatar   Among people working on virtual reality and cyberspace interfaces, an avatar is an icon or representation of a user in a shared virtual reality.

BBS   Bulletin Board Systems.

Blog   Short for a weblog, an online journal that can turn anyone with an internet connection into a mini-media outlet.

CMC   Computer Mediated Communication.

Cracker   One who breaks the security of a system. Coined in 1985 by hackers in defence against journalistic misuse of hacker.*

Cyberattacks   Attacks in cyberspace, such as defacements, worms and viruses.

Cyberconflict   Political conflict in computer-mediated environments. Takes two forms: ethnoreligious (two ethnic or religious groups fighting it out in cyberspace as they do in real life) and sociopolitical (social movements against antagonistic institutions).

Cyberdemocracy   The promise of nearly unlimited information delivered to your monitor in mere seconds is the promise of a better democracy.

Cyberdissidents   Dissidents fighting against authoritarian governments online using emails, file-trading, BBS and spamming e-magazines.

Cyberjihad or e-jihad   The online equivalent of Islamic fundamentalist jihad.

Cybernation   A nation whose communication of commonly held beliefs and philosophies is affected by the Internet or similar mechanisms.

Cyberpower   It is the form of power that structures culture and politics in cyberspace.

Cyberpunks   A movement or fashion trend associated especially with the rave/techno subculture.*

Cyberspace   Notional 'information space' loaded with visual cues and navigable with brain–computer interfaces.*

Cyberterrorism   Computer-based attacks intended to intimidate or coerce governments or societies in pursuit of goals that are political, religious or ideological.

Cyberwar   Refers to a more 'heavy' mode of new military conflict like destruction of the enemy's infrastructure through information technology.

C4ISR   Command, Control, Communications, Computing, Intelligence, Surveillance and Reconnaissance.

Defacement   The destruction of a web page on the internet.

Distributed Denial of Service attacks (DDoS)   DDoS attacks employ armies of 'zombie' machines (insecure server compromised by a hacker who places software on it that can launch an overwhelming number of requests, rendering the site inoperable).

Domain Name Service attacks (DNS)   Domain name servers are the 'Yellow Pages' that computers consult in order to obtain the mapping between the name of a system and the numerical address of the system. If the DNS provides an incorrect numerical adress for a website then the user's system will connect to the incorrect server. An attacker can disseminate false information this way and prevent access to the original site (Vatis 2001).

E-government   Government initiatives to provide online government services and civic information.

Email petitions   Petitions that contribute to electronic activism in their capacity for getting the word out about dates and times of organized protests, demonstrations and co-ordinated activities.

Flaming   To post an email message intended to insult and provoke.*

Gopher or archie servers   Gopher presents a menuing interface to a tree or graph of links; the link can be to documents, runnable programmes or other gopher menus arbitrarily far across the net.*

Hacker   Connotes a computer virtuoso who enjoys exploring the details of programmable systems and how to stretch their capabilities, one who programs enthusiastically, even obsessively.*

Hacktivism, online activism or electronic activism   Takes two forms: when hackers attack virtually chosen political targets and when persons organize through the internet or carry through email a political message.

ICT   Information and Communication Technologies

Information warfare or cyberwarfare   Sustained terrorist information warfare strategies are the ongoing deliberate efforts of an organized political group against the military, industrial, civilian and government economic information infrastructures or activities of a nation, region, organization of states, population or corporate entity (Erbschloe).

Internet filtering   Preventing viewers from viewing certain web content.

IRC   Internet Relay Chat.

ISP   Internet Service Provider.

IVDA   Individual Virtual Direct Action.

Malware   Malicious code programmes.

MVDA   Mass Virtual Direct Action.

Netwar   Refers to information age conflict at the less military, low intensity, more social end of the conflict's spectrum.

Network-centric warfare   Platforms are the eyes, ears and fists of a broader entity.

Ping saturation   Denial of service attacks where a target computer is overwhelmed with ping, a basic Internet program that lets you verify that a particular IP address exists and can accept requests.

Proxy server   A server that sits between a client application, such as a web browser and a real server. It is used to improve performance or filter requests. In cyberconflict, it allows a user to access blocked sites, not through a system controlled by the government.

Script kiddies   They commit mischief with scripts and programs written by others, often without understanding the exploit.

Spamming   To mass email unrequested identical email messages, particularly advertising; to crash a program by overwhelming a fixed size buffer with excessively large input data.*

SPIN   Segmented, polycentric, ideologically integrated network.

Swarming   Occurs when actors spread over great distances, and electronically converge on a target from multiple directions, a tactic different from the traditional form of attack in waves, which delivers a knockout blow from a single direction on the internet.

UNIX   Refers to the Unix operating system, an interactive timesharing system. Unix has become the most widely used multi-user general purpose operating system; and since 1996 the variant, Linux, has been at the cutting edge of the open source movement.*

USENET   From Users' Network – a distributed bulletin board system supported mainly by Unix machines.*

Virtual community or cybercommunity   An online community of people who have never met, but whose common beliefs move them to fight for, believe or even love their cybercommunity.

Virtual march   When protesters bombard the switchboard of an organization with calls and emails.

Virtual sit-in   Activists download a page aiming to flood the server.

Virus   Most viruses are worms, which are enabled by 'buffer overflows'. Buffer overflow is an event in which more data is put into a buffer (computer holding area) than the buffer has been allocated.

Notes

2 The three theories

1 Barry Zorthian, Chief Pentagon Public Affairs spokesman, during the Vietnam war, to a National Press Club forum on 19 March 1991.

4 Sociopolitical cyberconflicts

1 This SPIN concept, a precursor of the netwar concept, was proposed by Luther Gerlach and Virginia Hine in the 1960s to depict US social movements. It anticipates many points about network forms of organization that are now coming into focus in the analysis not only of social movements but also some terrorist, criminal, ethnonationalist and fundamentalist organizations. See Luther P. Gerlach (1987), 'Protest movements and the construction of risk', in B.B. Johnson and V.T. Covello (eds), *The Social and Cultural Construction of Risk*, Boston: D. Reidel Publishing Co., p. 115, based on Luther P. Gerlach and Virginia Hine (1970) *People, Power, Change: Movements of Social Transformation*, New York: The Bobbs-Merrill Co.
2 For examples of protest and hacktivist resources look at infoshop.org, crimethnic.com, broadleft.org, the hacktivist.com, electronic disturbance theatre, hackthissite.org, 2600.com, hbx.us, fromthewilderness.com, and an example of a counterconvention site www.counterconvention.org.
3 'True democracy, fake democracy or no democracy.' By Backgo, posted on the China Popular Marxist Liberal left-wing ideology website.

5 Ethnoreligious cyberconflict

1 For excellent excellent analysis of 9/11, al-Qaeda and the US response see Posner (2003), Friedman (2002) and Baudrillard (2002).

6 The effects of the internet on the 2003 Iraq war

1 C. Webb, 'Religious groups go online for peace', *Washington Post*. The article mentions: www.quaker.com, www.afsc.org (American friends service committee); Unitarian movement www.uua.org; Presbyterian church www.pcusa.org; World Council of Churches www.wcc-coe.org; the Washington-based group Churches for Middle East peace www.cmep.org, www.peaceprayer.com, American Jews www.peacenow.org, www.jewishpeacefellow.org, www.jewishpeacethread.com, www.peacelobby.org; Muslim sites www.ymca.org, www.icna.com, www.mpfweb.org.
2 For more on this and inter-media politics see G. Younge, 'Now dissent is "immoral"', *Guardian*, 02/06/03; D. Kennedy, 'The GOP attack machine: All who are not Bushies are evil', phoenix.com, 06/06/03; D. Lazare, 'The *New Yorker* goes to

war', *The Nation*, 02/06/03; B. Vann and D. North, 'Panic and hysteria reign at the *New York Times*', www.wsws.org; L. McQuaig, 'Bush unchallenged by media', *Toronto Star*, 25/05/03; the following articles from www.fair.org: S. Rendall, 'Dissent, disloyalty and double standards: Kosovo doves denounced Iraq war protest as anti-American', May/June 2003; J. Naureckas, 'Wolf Blitzer for the Defense (Department): Making sure the official line is the last word', January/February 2003; 'In Iraq crisis, networks are megaphones for official views', 18/03/03; N. Solomon, 'Media war: Obsessed with tactics and technology', 27/03/03.

3 For more analysis see S. Rendall and T. Broughel, 'Amplifying officials, squelching dissent: FAIR study finds democracy poorly served by war coverage', *Extra!*, May/June 2003. The news programmes studied were ABC World News Tonight, CBS Evening News, NBC Nightly News, CNN's Wolf Blitzer reports, Fox's Special Report with Brit Hume and PBS's Newshour with Jim Lehrer.

4 Also, there is an interesting quote on interactivity of the medium by Nigel Chapman, the deputy director of the BBC World Service in A. Lawson, 'War prompts text message boom', *Guardian*: 'Suddenly text messaging appears to have moved on from personal communication to personal statement ... New technologies are giving us a level of interaction with our audiences that we have never seen before.' According to the same article the volume of text messages to the broadcaster has grown tenfold since March to a total of 6,000. *Talking Point*, the show that enables listeners to quiz world leaders, has received more than 160,000 emails from listeners commenting on the war. The BBC Arabic version of *Talking Point* on its launch day received 3,000 emails.

5 M. Theodoulou, 'Proliferating Iranian weblogs give voice to taboo topics', *Christian Science Monitor*, 23/06/03; M. Dobbs, 'Iranian exiles sow change via satellite', *Washington Post*, 25/06/03; J. Curiel, 'North American media help Iran protests grow', *San Francisco Chronicle* 20/06/03; E. Batista, 'Bloggers report alternative news from G8', *Wired*, 04/06/03.

6 The Iraqi blogger was later revealed to be the interpreter of *Slate* journalist Peter Maas; see C. White, 'Iraqi blogger revealed', dot journalism, 12/06/03. Salam Pax's site featured in hundreds of news stories, including pieces by MSNBC, the BBC, the *New York Times* and the *Washington Post*. The bandwidth demands caused by the ensuing stampede of visitors overwhelmed servers, to the extent that Taylor Suchan, who runs Industrial Death Rock and Pyxz.com out of Texas, said he directed the links from the original photos to a parody image out of frustration, after trying contacting both Pax and Blogger, but received no response. The images on Suchan's servers are being viewed on average at least 140,000 times a day (M. Delio, 'Iraq blog: Hubbub over a headlock', *Wired*, 26/03/03).

7 The PewInternet survey on how online Americans use the internet and the Iraq war shows that blogs are gaining a following, but are not yet a source for the majority of users. Some 4 per cent of online Americans report going to blogs for information and opinions (www.pewinternet.org).

8 Users trying to log onto the al-Jazeera website in the US found a message that read 'hacked by Patriot, Freedom Cyberforce Militia', beneath the logo of the US flag (J. Deans, 'Hackers divert al-Jazeera users to US porn and patriot sites', *Guardian*, 28/03/03). According to Ballout, an employee of Network Solutions was tricked into giving the culprit a confidential password that allowed the hacker to temporarily assume total control of al-Jazeera's domain (M. Delio, 'Hackers condemn Arab site hack', *Wired*, 31/03/03). Despite this social engineering tactic, at the time it was also reported that the likely technique was DNS poisoning, which fools traffic-directing computers across the Internet, similar to vandalizing exit signs on an interstate to misdirect travellers. It is relatively difficult to defend against ('Hackers beat up on al-Jazeera', Associated Press, 27/03/03). Finally, in a plea agreement with the US Attorney's office, John William Racine, a 24-year-old web designer, admitted to

tricking VeriSign subsidiary Network Solutions into giving him ownership of the aljazeera.net domain. He turned himself in to FBI agents on 26 March, according to the plea agreement. He could have faced 25 years in prison, but if the judge agrees to the plea he will get three years of probation and 1,000 hours community service ('Al-Jazeera hacker admits guilt awaits sentence', www.silicon.com).

9 L. Walker, 'Casting a wider net for world news', *Washington Post*, 26/03/03, writes: 'Typical was this report in Australia's Age which said US media "had little hesitation in running graphical pictures of surrendering, captured, dead or dying Iraqi soldiers", and concluded: "It was a powerful insight into the enormous sway that the Bush Administration and the Pentagon exert over the media's coverage of the Iraq war."'

10 M. Delio, 'War worms inch across internet', 21/03/03, reports that at least three email viruses that their authors claim they were released in response to the war made rounds on the net. Virus writers often include messages tied to currents concerns like war, or eternal human urges like lust, to get people to open infected attachments. In widest circulation is Ganda, a low security threat by most security firms. Once attachments with references to the current military action are opened on PCs running Windows, Ganda behaves like many other email worms, emailing itself to all the addresses in the affected machine's Outlook contact list. It also scans the machine for security software and shuts them down.

11 According to the *New Hacker's Dictionary*, script kiddies do mischief with scripts and programs written by others, often without understanding the exploit; see M. Delio, 'Hackers condemn Arab site hacked', *Wired*, 31/03/03.

12 Kebt Lee, CEO of East View Cartographic, a firm that sells satellite-generated maps and other digital imagery, quoted in D. Tsuruoka, 'Internet, wireless to play key role in an Iraq war', yahoonews, 17/03/03. Also, Associated Press reported the possibility of the US military using 'e-bombs', which create a brief pulse of microwaves powerful enough to fry computers, blind radar, silence radios, and disable electronic ignitions in vehicles and aircraft. However, despite the e-bombs being classified, military analysts believe their range is a few yards at most ('E-bombs aims to stun Iraqi forces', Associated Press, 19/03/03).

Bibliography

Note: All the website addresses below were last accessed between 10 and 25 December 2005.

Abdel-Latif (December 2004) 'Cyber-struggle: Islamist websites versus the Egyptian state', *Arab Reform Bulletin*, Carnegie Endowment for International Peace. Online. Available at: www.carnegieendowment.org/files/PDF_December.pdf

Agence France Presse (14 August 2001) 'Chinese hackers attack Japanese websites over the Shrine visit'. Online. Available at: www.totse.com/en/politics/anarchism/162004.html

Alexander, A. (19 February 2003) 'Britain's biggest political protest was mobilised on the web', *Guardian*. Online. Available at: technology.guardian.co.uk/online/story/0,16545,898666,00.html

Aliefudien (30 December 2004) 'Al-Qaeda, big brother, and cyberspace', *Tehran Times*. Online. Available at: www.mehrnews.ir/en/NewsDetail.aspx?NewsID=142316

Allen, T. and Hudson, K. (eds) (1996) 'Introduction' in *War, Ethnicity and the Media*, London: South Bank University.

Always On (18 November 2003) 'Howard Dean blogs, Gen. Clark and President Bush follow', Always On. Online. Available at: www.alwayson-network.com/comments.php?id=A1297_0_5_0_C

American-Arab Anti-Discrimination Committee (13 November 2004) 'Protest biased media coverage of Palestine and Palestinians', ADC. Online. Available at: www.adc.org/index.php?id=2383

Amnesty International (26 November 2002) 'People's Republic of China: State control of the Internet in China', Amnesty International. Online. Available at: web.amnesty.org/ai.nsf/Index/asa170072002?OpenDocument&of=COUNTRIES%5C CHINA

Amnesty International (28 January 2004) 'China: Controls tighten as internet activism grows', Amnesty International Press Release. Online. Available at: web.amnesty.org/library/index/ENGASA170052004

Anderson, B. (1991) *Imagined Communities*, 2nd edn, London: Verso.

Anderson, J.W. (1998) *Arabizing the Internet*, The Emirates Occasional Papers, No. 30, The Emirates Center for Strategic Studies and Research.

Appadurai, A. (1996) *Modernity at Large: Cultural Dimensions of Globalization*, Minneapolis: University of Minnesota.

Arquilla, J. and Ronfeldt, D. (1997) *In Athena's Camp: Preparing for Conflict in the Information Age*, California: Rand.

Arquilla, J. and Ronfeldt, D. (2000) *Swarming and the Future of Conflict*, California: Rand.

Arquilla, J. and Ronfeldt, D. (eds) (2001) *Networks and Netwars: The Future of Terror, Crime and Militancy*, California: Rand.

Asaravala, A. (28 September 2002) 'College questioning site's link', Associated Press. Online. Available at: www.wired.com/news/politics/0,1283,55450,00.html

Asaravala, A. (16 December 2002) 'Treetop blogging protests logging', Wired.com. Online. Available at: www.wired.com/news/culture/0,1284,56660,00.html

Asaravala, A. (28 August 2003) 'Today's tech-dependent activists', Wired.com. Online. Available at: www.wired.com/news/technology/0,1282,60180,00.html

Ascribe (17 March 2003) 'Internet may mobilise largest protests ever seen, Professor predicts', Ascribe.org. Online. Available at: www.uwnews.org/article.asp?articleID=2088

Asian Pacific Post 'Why is Nortel helping China jail Internet users?', *Asian Pacific Post*. Online. Available at: www.asianpacificpost.com/news/article/136.html

Associated Press (30 May 2002) 'The war in all its online glory', Wired.com. Online. Available at: www.wired.com/news/politics/0,1283,52861,00.html

Associated Press (7 July 2002) 'Revised cybersecurity plan issued', Wired.com. Online. Available at: www.wired.com/news/conflict/0,2100,57109,00.html

Associated Press (27 August 2002) 'China dissidents thwarted on net', Wired.com. Online. Available at: www.wired.com/news/politics/0,1283,54789,00.html

Associated Press (15 October 2002) 'Finns investigate bomb chat room', Wired.com. Online. Available at: www.wired.com/news/politics/0,1283,55787,00.html

Associated Press (12 November 2002) 'Indictment due against hacker', Wired.com. Online. Available at: www.wired.com/news/politics/0,1283,56319,00.html

Associated Press (18 November 2002) 'Court approves more snooping', Wired.com. Online. Available at: www.wired.com/news/politics/0,1283,56454,00.html

Associated Press (20 December 2002) 'Frankfurt: nein to neo-Nazi sites', Wired.com. Online. Available at: www.wired.com/news/culture/0,1284,56945,00.html

Associated Press (7 January 2003) 'Revised cybersecurity plan issued', Wired.com. Online. Available at: www.wired.com/news/conflict/0,2100,57109,00.html

Associated Press (29 January 2003) 'Few clues in worm whodunit', Wired.com. Online. Available at: wired.com/news/infostructure/0,1377,57462,00.html

Associated Press (12 February 2003) 'Government warns "Patriotic hackers" against cyberattacks on Iraqi interests', Wired.com. Online. Available at: www.detnews.com/2003/technology/0302/14/technology-83639.htm

Associated Press (19 March 2003) 'E-bombs aims to stun Iraqi forces', Wired.com. Online. Available at: www.wired.com/news/conflict/0,2100,58122,00.html

Associated Press (27 March 2003) 'Hackers beat up on al-Jazeera', Wired.com. Online. Available at: www.wired.com/news/infostructure/0,1377,58238,00.html

Associated Press (9 April 2003) 'Internet fraud spikes sharply', Wired.com. Online. Available at: www.wired.com/news/culture/0,1284,58409,00.html

Associated Press (13 August 2003) 'Worm a sign of horrors to come', Wired.com. Online. Available at: www.wired.com/news/technology/0,1282,60019,00.html

Associated Press (21 August 2003) 'Ashcroft touts Patriot Act', Wired.com. Online. Available at: www.wired.com/news/conflict/0,2100,60102,00.html

Associated Press (27 August 2003) 'Is SCO hack a Linux attack?', Wired.com. Online. Available at: www.wired.com/news/business/0,1367,60303,00.html

Associated Press (29 August 2003) 'Suspected virus author arrested', Wired.com. Online. Available at: www.wired.com/news/technology/0,1282,60236,00.html

Associated Press (9 September 2003) 'Famed hacker turns himself in', Wired.com. Online. Available at: www.wired.com/news/conflict/0,2100,60102,00.html

Associated Press (12 September 2003) 'Are you too stupid to surf?', Wired.com. Online. Available at: www.wired.com/news/privacy/0,1848,60416,00.html

Associated Press (8 June 2004) 'Vietnam orders Net clampdown', Wired.com. Online. Available at: www.wired.com/news/print/0,1294,63764,00.html

Associated Press (25 September 2004) 'Google bows to Chinese censorship', Associated Press. Online. Available at: www.wired.com/news/politics/0,1283,65089,00. html?tw= newsletter_topstories_html

Associated Press (11 October 2004) 'US spies on chat rooms', Wired.com. Online. Available at: www.wired.com/news/privacy/0,1848,65305,00.html

Associated Press (14 April 2005) 'China's filters strong, subtle', Wired.com. Online. Available at: www.wired.com/news/privacy/0,1848,67221,00.html

Assur, A. (24 May 2004) 'Bloggers are outwitting the mainstream and corporate media', Uruk.net. Online. Available at: www.uruknet.info/?p=12028

Augman, R. (15 October 2004) 'The meaning behind the message of the RNC protests'. Online. Available at: groups.yahoo.com/groups/smygo/message/5343

Backgo (undated) posted on China Popular Marxist Liberal Left Wing Ideology website. Online. Available at: www.usembassy-china.org.cn/sandt/webdemocracy.html

Baker, P. (October 2001) 'Moral panic and alternative identity construction in Usenet', Department of Linguistics, Lancaster University. Online. Available at: www.ascusc.org/jcmc/vol7/issue1/baker.html

Barrett, N. (1996) *The State of the Cybernation*, London: Kogan Page.

Barthes, R. (1993) *Mythologies*, London: Vintage.

Barton, J. (22 June 2001) 'Ukraine's domain in dot dispute', Wired.com. Online. Available at: www.wired.com/news/politics/0,1283,44012,00.html

Batista, E. (4 June 2003) 'Bloggers report alternative news from G8', Wired.com. Online. Available at: www.wired.com/news/politics/0,1283,59086,00.html

Batista, E. and Dean, K.L. (21 August 2003) 'Vague limits vex music traders', Wired.com. Online. Available at: www.wired.com/news/mp3/0,1285,60110,00. html

Baudrillard, J. (1983) *Simulations*, New York: Semiotext.

Baudrillard, J. (1988) *The Ecstasy of Communication*, NewYork: Semiotext.

Baudrillard, J. (1988) *Selected Writings*, ed. Mark Poster, Cambridge: Polity Press.

Baudrillard, J. (1993) *The Transparency of Evil*, New York: Verso.

Baudrillard, J. (1995) *The Gulf War Did Not Take Place*, Bloomington: Indiana University Press.

Baudrillard, J. (2003) *The Spirit of Terrorism and Other Essays*, trans. Chris Turner, London and New York: Verso.

BBC News (16 October 1998) 'Indian army website ambushed'. Online. Available at: http://news.bbc.co.uk/hi/english/world/south_asia/newsid_194000/194844.stm

BBC News (25 October 1998) 'War of words on the Internet'. Online. Available at: http://news.bbc.co.uk/hi/english/world/monitoring/newsid_200000/200708.stm

BBC News (14 April 2000) 'Serb hackers on the rampage'. Online. Available at: http://news.bbc.co.uk/hi/english/world/europe/newsid_712000/712211.stm

BBC News (3 November 2000) 'Israeli group hacked'. Online. Available at: http://news.bbc.co.uk/hi/english/world/middle_east/newsid_1005000/1005850.stm

BBC News (16 April 2002) 'Israel under hack attack'. Online. Available at: http://news.bbc.co.uk/hi/english/sci/tech/newsid_1932000/1932750.stm

BBC News (23 September 2002) 'The cost of China's web censors'. Online. Available at: http://news.bbc.co.uk/2/hi/business/2264508.stm

BBC News (4 December 2002) 'China blocks news not porn online'. Online. Available at: http://news.bbc.co.uk/2/hi/technology/2540309.stm

BBC News (29 June 2003) 'China jails web dissidents'. Online. Available at: http://news.bbc.co.uk/1/hi/world/asia-pacific/2946526.stm

BBC News (22 July 2003) 'China's web surfers keep growing'. Online. Available at: http://news.bbc.co.uk/2/hi/technology/3086391.stm

BBC News (15 August 2003) 'China censors online chat'. Online. Available at: http://news.bbc.co.uk/1/hi/technology/3034023.stm

BBC News (8 September 2003) 'Computer worm targets Blair'. Online. Available at: http://news.bbc.co.uk/2/hi/technology/3083192.stm

BBC News (24 September 2003) 'China online dissident charged'. Online. Available at: http://news.bbc.co.uk/1/hi/world/asia-pacific/3134306.stm

BBC News (29 September 2003) 'Discord at digital divide talks'. Online. Available at: http://news.bbc.co.uk/1/hi/technology/3148356.stm

BBC News: (29 October 2003) 'Islamic hackers step up attacks'. Online. Available at: http://news.bbc.co.uk/1/hi/technology/2372209.stm

BBC News (3 November 2004) 'Bypassing China's net firewall'. Online. Available at: http://news.bbc.co.uk/2/hi/technology/3548035.stm

BBC News (16 June 2005) 'Chinese blogs face restrictions'. Online. Available at: http://news.bbc.co.uk/2/hi/technology/4617657.stm

BBC News (7 July 2005) 'Blogs respond to London blasts'. Online. Available at: http://news.bbc.co.uk/1/hi/technology/4659679.stm

Bell, D. and Kennedy, B. (eds) (2000) *The Cybercultures Reader*, London and New York: Routledge.

Bell, E. (30 March 2003) 'The blogs of war', *Observer*.

Bennett, W.L. (2004) 'Communicating global activism' in Van de Donk, W., Loader, B., Nixon, P. and Rucht, D. (eds), *Cyberprotest: New Media, Citizens and Social Movements*, London and New York: Routledge.

Berton, P., Kimura, H., Zartman, I.W. (eds) (1999) *International Negotiation*, New York: St. Martin's Press.

Beyerle, S. (July/August 2002) 'Net effect: The Middle East's e-war', *Foreign Policy*. Online. Available at: www.foreignpolicy.com/story/cms.php?story_id=177

Bhabha, H. (1994) *The Location of Culture*, New York: Routledge.

Bloom, W. (1990) *Personal Identity, National Identity and International Relations*, Cambridge: Cambridge University Press.

Bodeen, C. (3 September 2002) 'Search engine Google blocked in China', Associated Press. Online. Available at: www.ctv.ca/servlet/ArticleNews/story/CTVNews/1096138932491_76/?hub=SciTech

Borger, J. (9 November 1999) 'Pentagon kept lid on cyberwar in Kosovo', *Washington Guardian, Guardian Unlimited*. Online. Available at: www.guardian.co.uk/Kosovo/Story/0,2763,197391,00.html

Boucher, J., Landis, D. and Clarke, A.K. (eds) (1987) *Ethnic Conflict: International Perspectives*, London and New York: Sage.

Bowles, W. (26 September 2004) 'Embedded in a media fantasy world: The case against objective reporting'. Online. Available at: www.williambowles.info/ini/ini-0273.html

Bowles, W. (6 March 2005) 'Privatising propaganda'. Online. Available at: www.williambowles.info/ini/ini-0312.html

Boyle, J. (1997) 'Foucault in cyberspace: surveillance, sovereignty and hard-wired censors', Duke University. Online. Available at: www.law.duke.edu/boylesite/foucault.htm

Broadbendt, L. (1985) *War and Peace News*, Glasgow University Media Group, Milton Keynes: Open University Press.

Brown, R. (February 1999) 'Conceptualizing the impact of communication technologies on world politics: space, time, and mobilization', paper presented at the International Studies Association Conference, Washington DC.

Brown, R. (August/September 2000) 'Mobilizing the bias of communication: information technology', Political Communications and Transnational Political Strategy, paper presented at the American Political Science Association Convention, Washington DC.

Brown, R. (2002) 'The contagiousness of conflict: E.E. Schattschneider as a theorist of the information society', *Information, Communication and Society*, 5: 2.

Browning, G. (1996) *Electronic Democracy*, London: Wilton.

Brint, S. (1984) 'New class and cumulative trend explanations of the liberal attitudes of professionals', *American Journal of Sociology*, 90: 30–71.

Burton, J. (1997) *Violence Explained: The Sources of Conflict, Violence and Crime and their Prevention*, New York: Manchester University Press.

Butko, T.J. (May 2004) 'Revelation or revolution: A Gramscian approach to the rise of political Islam', *British Journal of Middle Eastern Studies*, 31(1). Online. Available at: http://taylorandfrancis.metapress.com/index/V9WJF0QEQLLR92T3.pdf

Cairns, E. (1997) *A Safer Future*, England: Oxfam Publications.

Callinicos, A. (July 2002) 'Regroupment, realignment and the Revolutionary Left'. Online. Available at: www.swp.org.uk/INTER/regroupmen.pdf

Campbell, D. (1992) *Writing Security*, Manchester: Manchester University Press.

Campbell, D. (1998) *National Reconstruction: Violence, Identity and Justice in Bosnia*, Minneapolis: University of Minnesota Press.

Campbell, D. (1999) 'Violence, justice and identity' in Edkins, J., Persram, N. and Pin-Fat, V. (eds) *Sovereignty and Subjectivity*, Boulder and London: Lynne Pienner.

Castells, M. (2000) *The Rise of the Network Society*, The Information Age: Economy, Society and Culture, vol. 1, 2nd edn, Oxford: Blackwell.

CBS News (25 July 2003) 'Video of Saddam's sons released'. Online. Available at: www.cbsnews.com/stories/2003/07/22/iraq/main564599.shtml

Cebrian, J.L. (2000) *La Red (The Net)*, translated in Greek, Athens: Stahy.

Chanteur, J. (1992) *From War to Peace*, Boulder: Westview Press.

Chapman, R. (undated) 'Hyperlinks and HyperProtestantism: The Internet as a postmodern epistemological shift', Robinson, Illinois: Lincoln Trail College.

Chase, M. and Mulveron, J. (2002) *You've Got Dissent! Chinese Dissident Use of the Internet and Beijing's Counterstrategies*, California: Rand.

Chayko, M. (2002) *Connecting: How We Form Social Bonds and Communities in the Internet Age*, New York: State University of New York Press.

Chepsiuk, R. (2001) 'Get ready for cyberwars'. Online. Available at: www.afcea.org.ar/publicaciones/chepesiuk.htm

Chernaik, W., Deegan, M. and Gibson, A. (1996) *Beyond the Book: Theory, Culture, and the Politics of Cyberspace*, Office for Humanities Communication Publications, no. 7, The Centre for English Studies, University of London.

Chesters, G. (2003) 'Shape shifting: Civil society, complexity and social movements', *Anarchist Studies*, 11(1): 42–65.

Chesters, G. (22 October 2004) 'ESF: encounter or representation', Indymedia.org. Online. Available at: www.nadir.org/nadir/initiativ/agp/free/wsf/london2004/1022encounter.htm

Ching, F. (22 May 2003) 'The Internet dissidents', *Globe and Mail*. Online. Available at: www.theglobeandmail.com

Cleaver, H. (1999) 'Computer-linked social movements and the global threat to capitalism'. Online. Available at: www.eco.utexas.edu/faculty/Cleaver/hmchtmlpapers.html

Clemens, E. (1996) 'Organizational form as frame: Collective identity and political strategy in the American labor movement, 1880–1920', in McAdam, D., McCarthy, J. and Zald, M. (eds), *Comparative Perspectives on Social Movements: Political Opportunities, Mobilizing Structures, and Structural Framings*, Cambridge: Cambridge University Press.

CNETAsia (19 June 2002) 'Experts warn of cyber security holes'. Online. Available at: www.net-security.org/news.php?id=425

CNETAsia (2 September 2003) 'Malaysians enrol text messages for democracy', CNETAsia. Online. Available at: news.zdnet.co.uk/communications/wireless/ 0,39020348,39116064,00.html

Cohen, J.L. (1985) 'Strategy or identity: New theoretical paradigms and contemporary social movements', *Sociological Research*, 52: 663–716.

Coll, S. (29 March 2005) 'In the Gulf, dissidence goes digital: Text messaging is new tool of political underground', *Washington Post* Foreign Service. Online. Available at: www.washingtonpost.com/wp-dyn/articles/A8175-2005Mar28.html

Collins, M. (Fall 1992) 'Flaming: The relationship between social context cues and uninhibited verbal behavior in computer-mediated communication'. Online. Available at: www.emoderaters.com/papers/flames.html

Cornell, S. and Hartmann, D. (1998) *Ethnicity and Race: Making Identities in a Changing World*, Thousand Oaks, CA: Sage.

Costello, S. (5 December 2002) 'Attrition.com stops mirroring hacked websites', CNN.com. Online. Available at: http://edition.cnn.com/2001/TECH/internet/05/23/ attrition.mirroring.idg

Cramer, C. (9 March 2003) 'How the net will play a key role in this war', *Observer*. Online. Available at: www.poynter.org/column.asp?id=60&aid=24305

Crandall, J. (15 June 1999) 'Anything that moves: Armed vision'. Online. Available at: www.ctheory.net/articles.aspx?id=115

Cronauer, K. (2001) 'Activism and the Internet: Uses of electronic mailing lists by social activists', University of Columbia. Online. Available at: www.ssrc.org/programs/itic/ publications/civsocandgov/cronaueracademic.pdf

Cryan, P. (1 December 2004) 'Rightist bias in wire coverage of Colombia', Counterpunch.org. Online. Available at: www.counterpunch.org/cryan12012004.html

Crypt Newsletter (1999) Online. Available at: www.soci.niu.edu/~crypt/other/hamrewat.html

Curiel, J. (20 June 2003) 'North American media help Iran protests grow', *San Francisco Chronicle*. Online. Available at: www.sfgate.com/cgi-bin/article.cgi?file=/chronicle/ archive/2003/06/20/MN293330.DTL

Dachan, M. (undated) *Internet Usage in the Middle East: Some Political and Social Implications*, Department of Political Science, The Hebrew University of Jerusalem.

David, M.W. and Sakurai, K. (2003) 'Combating Cyberterrorism: Improving analysis and accountability', *Journal of Information Warfare*, 2: 15–26.

Davis, J., Hirschl, T. and Stack, M. (1997) *Cutting Edge*, London and New York: Verso.

Davis, R. (1999) *The Web of Politics*, Oxford: Oxford University Press.

De Armond, P. (2001) 'Netwar in the Emerald City: WTO protest strategy and tactics' in Arquilla, J. and Ronfeldt, D. (eds), *Networks and Netwars: The Future of Terror, Crime and Militancy*, California: Rand.

De Fleur, M. and Ball-Roceach, S. (1982) *Theories of Mass-communication*, 4th edn, New York and London: Longman.

De Kerckhove, D. (1997) *Connected Intelligence: The Arrival of the Web Society*, London: Kogan Page.

Dean, K. (1 August 2003) 'Schools rebuke music biz demands', Wired.com. Online. Available at: www.wired.com/news/digiwood/0,1412,59726,00.html

Dean, K. (26 April 2004) 'Online anonymity comes under fire', Wired.com. Online. Available at: www.wired.com/news/digiwood/0,1412,58633,00.html

Deans, J. (28 March 2003) 'Hackers divert al-Jazeera users to US porn and patriot sites', *Guardian*. Online. Available at: www.guardian.co.uk/Iraq/Story/0,2763,924905,00.html

Deleuze, G. and Guattari, F. (1987) *A Thousand Plateaus*, London: The Athlone Press.

Delio, M. (30 April 2001) 'It's Cyberwar: China vs US', Wired.com. Online. Available at: www.wired.com/news/politics/0,1283,43437,00.html

Delio, M. (29 June 2002) 'Attrition offs its hacker monitor', Wired.com. Online. Available at: www.wired.com/news/culture/0,1284,43991,00.html

Delio, M. (30 August 2002) 'Did FBI bungle email evidence?', Wired.com. Online. Available at: www.wired.com/news/conflict/0,2100,54857,00.html

Delio, M. (18 October 2002) 'Terror turns real for horror site', Wired.com. Online. Available at: www.wired.com/news/technology/0,1282,55848,00.html

Delio, M. (19 November 2002) 'How much hack info is too much?', Wired.com. Online. Available at: www.wired.com/news/infostructure/0,1377,56463,00.html

Delio, M. (26 November 2002) 'Cyber-rights activists log a win', Wired.com. Online. Available at: www.wired.com/news/politics/0,1283,56577,00.html

Delio, M. (16 December 2002) 'Raided firm's software checks out', Wired.com. Online. Available at: www.wired.com/news/conflict/0,2100,56777,00.html

Delio, M. (30 December 2002) 'So many holes, so few hacks', Wired.com. Online. Available at: www.wired.com/news/infostructure/0,1377,56955,00.html

Delio, M. 'What Symantec knew but didn't say', Wired.com. Online. Available at: www.wired.com/news/infostructure/0,1377,57676,00.html

Delio, M. (13 March 2003) 'Yaha Virus uses netizens as pawns', Wired.com. Online. Available at: www.wirednews.com/news/infostructure/0,1377,58026,00.html

Delio, M. (21 March 2003) 'War worms inch across internet', Wired.com. Online. Available at: www.wired.com/news/infostructure/0,1377,58143,00.html

Delio, M. (26 March 2003) 'Iraq blog: Hubbub over a headlock', Wired.com. Online. Available at: www.wired.com/news/culture/0,1284,58206,00.html

Delio, M. (28 March 2003) 'Blogs opening Iranian society?', Wired.com. Online. Available at: www.wired.com/news/culture/0,1284,58976,00.html

Delio, M. (31 March 2003) 'Hackers condemn Arab site hack', Wired.com. Online. Available at: www.wired.com/news/infostructure/0,1377,58277,00.html

Delio, M. (3 April 2003) 'Do privacy fears allow terrorism?', Wired.com. Online. Available at: www.wired.com/news/privacy/0,1848,58332,00.html

Delio, M. (13 August 2003) 'Worm exploits weak link: PC users', Wired.com. Online. Available at: www.wired.com/news/infostructure/0,1377,59994,00.html

Delio, M. (30 August 2003) 'Blaster worm still making mayhem', Wired.com. Online. Available at: www.wired.com/news/infostructure/0,1377,60237,00.html

Delio, M. (22 August 2003) 'This worm ain't gonna hunt', Wired.com. Online. Available at: www.wired.com/news/infostructure/0,1377,60150,00.html

Delio, M. (28 January 2004) 'Worm slowing, but still dangerous', Wired.com. Online. Available at: www.wired.com/news/technology/0,1282,62073,00.html

Delio, M. (2 September 2003) 'Net analysis gets turbo boost', Wired.com. Online. Available at: www.wired.com/news/infostructure/0,1377,60077,00.html

Delio, M. (12 September 2003) 'Manhattan mob meets its maker', Wired.com. Online. Available at: www.wired.com/news/culture/0,1284,60399,00.html

Delio, M. (15 September 2003) 'Flash mobs get a dash of danger', Wired.com. Online. Available at: www.wired.com/news/culture/0,1284,60364,00.html

Delio, M. (15 September 2003) 'Viruses, worms: what's in a name?', Wired.com. Online. Available at: www.wired.com/news/infostructure/0,1377,60281,00.html

Delio, M. (11 November 2003) 'The Internet is a very sick place', Wired.com. Online. Available at: www.wired.com/news/infostructure/0,1377,61710,00.html

Delio, M. (12 July 2004) 'A gathering to hack the system', Wired.com. Online. Available at: www.wired.com/news/infostructure/0,1377,64172,00.html

Delio, M. (12 July 2004) 'Hactivism and how it got there', Wired.com. Online. Available at: www.indymedia.org.uk/en/2004/10/299962.html

Denning, D. (2001) 'Is cyberterrorism next?', Wired.com. Online. Available at: www.ssrc.org/sep11/essays/denning/html

Denning, D. (2001) 'Activism, hacktivism and cyberterrorism: The Internet as a tool for influencing foreign policy' in Arquilla, J. and Ronfeldt, D. (eds), *Networks and Netwars: The Future of Terror, Crime and Militancy*, California: Rand.

Denning, D. and Baugh, W.E. (2000) in Thomas, D. and Loader, B. (eds) *Cybercrime: Law Enfrocement and Surveillance in the Information Age*, London: Routledge.

Dery, M. (1993) *Flame Wars: The Discourse of Cyberculture*, North Carolina: Duke University Press.

Diamond, L. and Plattner, M. (1994) *Nationalism, Ethnic Conflict and Democracy*, Baltimore and London: Johns Hopkins University Press.

Diani, M. (16–17 September 1999) 'Social Movement networks virtual and real', paper presented at conference 'A New Politics?' CCSS, University of Birmingham. Online. Available at: www.nd.edu/~dmyers/cbsm/vol2/bgham99.pdf

Dobbs, M. (25 June 2003) 'Iranian exiles sow change via satellite', *Washington Post*. Online. Available at: www.dailyalert.org/archive/2003-06/2003-06-27.html

Dodge, M. and Kitchin, R. (2001) *Mapping Cyberspace*, London and New York: Routledge.

Dot Journalism (8 September 2003) 'Al-Jazeera launches English-language site', Journalism.co.uk. Online. Available at: www.journalism.co.uk/news/story709.shtml

Dot Journalism (18 November 2003) 'Zimbabwe's daily news battles online', Dot Journalism. Online. Available at: www.journalism.co.uk/news/story727.html

Economist, The (21 January 2006) 'Old mogul, new media'.

Eder, K. (1985) 'The "new" social movements: moral crusades, political pressure groups or social movements?', *Sociological Research*, 52: 869–901.

Edwards, A. (2004) 'The Dutch women's movement online' in Van de Donk, W., Loader, B., Nixon, P. and Rucht, D. (eds), *Cyberprotest: New Media, Citizens and Social Movements*, London and New York: Routledge.

Elliot, M. and Scacchi, W. (August 2004) 'Mobilization of software developers: the Free Software Movement'. Online. Available at: http://opensource.mit.edu/papers/elliottscacchi2.pdf

Emery, E. (undated) 'The postmodern desert: Solitude and community in cyberspace', Preconvention papers, The American Benedictine Academy. Online. Available at: www.osb.org/aba/aba2000/emery.html

Europemedia.net (1 April 2003) 'Iraq invasion leads to massive increase in pro-war, anti-

war hacker activity', Europemedia.net. Online. Available at: www.internetsecuri-
tynews.com/blogger/2003/04/europemedia

Europemedia.net (14 April 2003) '"Super-network" will combine internet, fixed-line and
mobile'. Online. Available at: www.allbusiness.com/periodicals/issue/68733-1-2.html

Europemedia.net (19 May 2003) 'Dutch professor calls for openness in IT security',
Europemedia.net. Online. Available at: www.allbusiness.com/periodicals/issue/70520-
1-2.html

Europemedia.net (9 July 2003) 'New website gets behind municipal wireless projects',
Europemedia.net. Online. Available at: www.egov.vic.gov.au/Research/WAP/wap.htm

Everard, J. (2000) *Virtual States*, London and New York: Routledge.

Fair.org (3 April 2003) 'Some critical media voices face censorship', Fair and Accuracy
in Reporting, Media Advisory. Online. Available at: www.fair.org/press-releases/iraq-
censorship.html

Fair.org (4 April 2003) 'Journalist's evidence that US bombed market ignored by US
press', Fair and Accuracy in Reporting, Media Advisory. Online. Available at:
www.fair.org/index.php?page=1613

Fair.org (12 March 2004) 'The return of PSYOPS: Military's media manipulation
demands more investigation', Fair and Accuracy in Reporting, Media Advisory.
Online. Available at: www.fair.org/index.php?page=1983

Fairclough, N. (1995) *Media Discourse*, London and New York: Arnold.

Farley, M. 'Dissidents hack holes in China's new wall', *Los Angeles Times*. Online.
Available at: www.gis.net/~cht/dissidents.html

Ferdinard, P. (ed.) (2000) *The Internet, Democracy and Democratization*, London and
Portland: Frank Cass.

Fink, J. (1999) *Cyberseduction: Reality in the Age of Psychotechnology*, New York:
Prometheus.

First Monday 'LiveJournal as site of knowledge creation and sharing', First Monday.
Online. Available at: www.firstmonday.org/issues/issue9_12/raynes/

Foucault, M. (1979) *Discipline and Punish*, London: Prentice Hall.

Foucault, M. (1980) *Power/Knowledge*, New York and London: Harvester and Wheat-
sheaf.

Franda, M. (2002) *Internet Development and Politics in Five World Regions*, Boulder:
Lynne Rienner.

Freeden, M. (1996) *Ideologies and Political Theory: A Conceptual Approach*, Oxford:
Clarendon Press.

Freidman, T. (2003) *Longitudes and Attitudes: Exploring the World Before and After
September 11*, London: Penguin.

Frishberg, M. (27 May 2002) 'Local access: IT takes a village', Wired.com. Online.
Available at: www.gis.net/~cht/dissidents.html,   www.wired.com/news/culture/
0,1284,52690,00.html

Frishberg, M. (21 July 2003) 'Roll-your-own net TV takes off', Wired.com. Online.
Available at: www.wired.com/news/technology/0,1282,59623,00.html

Frissen P. (1997) 'The virtual state: postmodernization, informatization, and public
administration' in Loader, B. (ed.), *The Governance of Cyberspace*, London: Routledge.

Gadet, F. (1986) *Sassure and Contemporary Culture*, London: Hutchinson.

Galtung, J. *Peace by Peaceful Means: Peace and Conflict, Development and Civilisation*,
London and New York: Prio, Sage.

Gamson, W. and Meyer, D. (1996) 'Framing political opportunity' in McAdam, D,

McCarthy, J. and Zald, M. (eds), *Comparative Perspectives on Social Movements: Political Opportunities, Mobilizing Structures, and Structural Framings*, Cambridge: Cambridge University Press.

Gardner, J. (11 September 2002) 'Terror czar: the war is digital', Wired.com. Online. Available at: www.wired.com/news/politics/0,1283,55089,00.html

Gentile, C. (8 November 2000) 'Hacker war rages in Holy land', Wired.com. Online. Available at: www.wired.com/news/politics/0,1283,40030,00.html

Georgiou, M. (October 2002) 'Diasporic communities on-line: a bottom up experience of transnationalism', *Hommes et Migrations*.

Gerodimos, R. (2004) 'Social movements and online civic engagement', paper given at 2nd International Conference Imaging Social Movements, Edge Hill College, 1–3 July.

Gibson, R., Nixon, P. and Ward, S. (2003) *Political Parties and the Internet: Net Gain?*, London and New York: Routledge.

Gittings, J. (17 September 2002) 'Sense and censorship', *Guardian*. Online. Available at: www.guardian.co.uk/china/story/0,7369,793813,00.html

Glasner, J. (10 October 2002) 'Mobile junkies reshaping society', Wired.com. Online. Available at: www.wired.com/news/culture/0,1284,55561,00.html

Glasner, J. (19 November 2002) 'A vote for less tech at the polls', Wired.com. Online. Available at: www.wired.com/news/business/0,1367,56370,00.html

Glasner, J. (19 March 2003) 'Protests to start when war does', Online. Available at: www.wired.com/news/politics/0,1283,58101,00.html

Glasner, J. (16 April 2003) 'Security biz thrives on fear', Wired.com. Online. Available at: www.wired.com/news/infostructure/0,1377,58492,00.html

Glasner, J. (30 May 2003) 'Media more diverse? Not really', Wired news. Online. Available at: www.wired.com/news/business/0,1367,59015,00.html

Glasner, J. (6 August 2003) 'Patriot Act legal attacks pile up', Wired.com. Online. Available at: www.wired.com/news/business/0,1367,59863,00.html

Glasner, J. (17 August 2003) 'Blackout a boon for backup tools', Wired.com. Online. Available at: www.wired.com/news/business/0,1367,60048,00.html

Glave, G. (8 January 1999) 'Crackers call off war', Wired.com. Online. Available at: www.wired.com/news/news/technology/story/17231.html

Globe Technology (10 May 2001) 'Chinese hackers halt webwar, say 1,000 US sites defaced'. Online. Available at: www.globeandmail.com

Gorr, W. (September 2003) 'Cloudy, with a chance of theft', Wired.com: 11. Online. Available at: www.wired.com/wired/archive/11.09/view.html

Graham, B. (7 February 2003) 'Bush orders guidelines for cyberwarfare', *Washington Post*. Online. Available at: www.washingtonpost.com/wp-srv/liveonline/03/special/technews/sp_technews_graham020703.htm

Grebb, M. (3 October 2002) 'Windows, Unix still at risk', Wired.com. Online. Available at: www.wired.com/news/politics/0,1283,55544,00.html

Greenspan, R. (18 July 2003) 'Education: reading, writing, pointing-and-clicking'. Online. Available at: www.clickz.com/stats/sectors/education/article.php/2237481

Greimel, H. (26 August 2002) 'Psychiatry group hopes to probe Chinese mental wards for dissidents', Associated Press. Online. Available at: back.faluninfo.net/displayAnArticle.asp?ID=6182

Grewlich, K. (1999) *Governance in 'Cyberspace'*, The Hague, London and Boston: Kluwer Law International.

Guardian Unlimited (26 June 2002) 'Vietnam to monitor "reactionary" web use'. Online. Available at: www.guardian.co.uk/freespeech/article/0,2763,861190,00.html

Guisnel, J. (1997) *Cyberwars*, New York and London: Plenum Trade.

Gumbel, A. (2003) (16 March 2003) 'Global peace movement is alive and kicking', *Independent on Sunday*. Online. Available at: www.iol.co.za/index.php?sf=2813&set_id=&sf=2813&click_id=3&art_id=ct20030316200614173W6163448&set_id=1

Gumbel, A. (2 February 2006) 'Google shares slump after profits disappoint market', *Independent*.

Hadar, L. (12 March 2005) 'Pseudo-events stir Mideast pot'. Online. Available at: www.antiwar.com/orig/hadar.php?articleid=5166

Hall, S. (1997) 'The local and the global' in McClintock, A. et al. (eds), *Dangerous Liaisons: Gender, Nation and Postcolonial Perspectives*, Minneapolis: University of Minnesota.

Harding, D. (10 June 2004) 'Vigilantes fight back by hacking the hackers', *Metro*.

Hardt, M. and Negri, A. (2000) *Empire*, Cambridge, MA and London: Harvard University Press.

Hardt, M. and Negri, A. (2004) *Multitude*, New York: Penguin Press.

Hasson, J. (8 September 2003) 'Surprising percentage of public fear cyberattacks', usatoday.com. Online. Available at: www.usatoday.com/tech/news/2003-08-30-cyberterror_x.htm

Heim, M. (1993) *The Metaphysics of Virtual Reality*, Oxford: Oxford University Press.

Held, D. (1995) *Democracy and the Global Order*, Cambridge: Polity Press.

Held, D. and McGrew, A. (2000) *The Global Transformation Reader*, Cambridge: Polity Press.

Held, D. and McGrew, A. (2000) 'The great globalization debate: An introduction', in Held, D. and McGrew, A. (eds), *The Global Transformation Reader*, Cambridge: Polity Press.

Hermida, A. (6 May 2002) 'Palestinian websites knocked offline', BBC News. Online. Available at: http://news.bbc.co.uk/2/hi/science/nature/1966335.stm

Hershman, T. (29 June 2001) 'Israel discusses "Inter-Fada"', Wired.com. Online. Available at: www.wired.com/news/politics/0,1283,41154,00.html

Heskett, B. (5 February 2002) 'Who's benefiting from Net attacks?' Online. Available at: www.zdnet.com/zdnn.news.zdnet.com/2100-1009_22-829686.html

Hill, K. and Hughes, J. (1998) *Cyberpolitics*, Lanham: Rowman and Litttlefield.

Hindess, B. (1999) *Discourses on Power: From Hobbes to Foucault*, Oxford: Blackwell.

Higgins, J. (1997) 'Peace Profile: Paulo Freire', *Peace Review*, 9(4): 571–577.

Hines, M. (22 October 2004) 'Paper's anti-Bush ploy gets hacked, sacked', CNET News.com. Online. Available at: http://news.com.com/UK+papers+anti-Bush+ploy+gets+hacked,+sacked/2100-1028_3-5423204.html

Hintjens, H. 'When identity becomes a knife: Reflecting on the genocide in Rwanda', *Ethnicities*, 1(1): 25–55.

Hockstader, L. (27 October 2000) 'Pings and e-arrows fly in Mideast cyber-war', *Washington Post*. Online. Available at: www.washingtonpost.com/ac2/wp-dyn?pagename=article&node=&contentId=A21154-2000Oct26

Hoffman, B. (1998) *Inside Terrorism*, London: Victor Gollancz.

Hood, Robin (29 August 2004) 'Hacktivists use corporate credit cards to donate to humanitarian and civil liberties groups', Indymedia.org. Online. Available at: http://brasil.indymedia.org/en/blue/2004/08/289523.shtml

Hopper, I. (20 March 2000) 'Kashmir conflict continues to escalate online', CNN.com. Online. Available at: www.cnn.com/2000/TECH/computing/03/20/pakistani.hackers

Horowitz, D. (1998) 'Structure and strategy in ethnic conflict', paper presented at Annual Conference on Development Economics, Washington DC, 20–21 April.

Howarth, D. (2000) *Discourse*, Buckingham and Philadelphia: Open University Press.

Howarth, D., Norval, J. and Stavrakakis, Y. (eds) (2000) *Discourse Theory and Political Analysis: Identities, Hegemonies and Social Change*, Manchester and New York: Manchester University Press.

IFEX (18 November 2003) 'New wave of Chinese dissident arrests', Information Society News. Online. Available at: www.ifex.org/en/content/view/full/13242

Ignatieff, M. (2000) *Virtual War: Kosovo and Beyond*, London: Chatto and Windus.

Indymedia (7 January 2004) 'FBI took the hard drives of IMC servers in the UK', Press-release, IMC. Indymedia. Online. Available at: www.indymedia.org/en/2004/10/111987.shtml

Indymedia (5 September 2004) 'Hacktivism: A week of electronic disruption during the Republican National Convention', Indymedia. Online. Available at: http://colorado.indymedia.org/newswire/display/8890/index.php

Indymedia (15 December 2004) 'CIA behind automated chat room spying scheme', Indymedia. Online. Available at: www.indymedia.org.uk/en/2004/12/302852.html

Indymedia (10 January 2005) 'Electronic civil disobedience and solidarity for Gelmart workers', www.indymedia.org. Online. Available at: http://manila.perthimc.asn.au/?action=default&featureview=113

Jabri, V. (1996) *Discourses on Violence: Conflict Analysis Reconsidered*, Manchester: Manchester University Press.

Jamail, D. (14 November 2004) 'Media repression in "Liberated" Land'. Online. Available at: www.dahrjamailiraq.com/hard_news/archives/2004_11_14.php

Jame, F. (10 August 1999) 'China, Taiwan in web hacking "war"', ZDNet. Online. Available at: http://news.zdnet.com/2100-9595_22-515403.html

Jardin, X. (29 October 2002) 'P2P App's aim: defend free speech', Wired.com. Online. Available at: www.wired.com/news/technology/0,1282,56063,00.html

Jenkins, J.C. (1983) 'Resource mobilization theory and the study of social movements', *Annual Review of Sociology*, 9: 527–553.

Jesdanun, A. (13 August 2003) 'Computer infection snarls global network', newsday.com. Online. Available at: www.eweek.com/article2/0,1895,1655952,00.asp

Jin, R. (11 July 2005) 'Korea: Roh adopts e-mail politics', *Korea Times*. Online. Available at: www.asiamedia.ucla.edu/article.asp?parentid=30957

Joinson, A. (2002) *Understanding the Psychology of Internet Behaviour*, Palgrave: Macmillan.

Jones, S. (ed.) (1999) *Doing Internet Research*, Thousand Oaks, London and New Delhi: Sage.

Joo-hee, L. (11 February 2005) 'Korea: Internet a key playing field for politicians, but has pitfalls', *Korea Herald*. Online. Available at: www.asiamedia.ucla.edu/article.asp?parentid=20688

Jordan, T. (1999) *Cyberpower: The Culture and Politics of Cyberspace and the Internet*, London and New York: Routledge.

Jordan, T. (2001) 'Mapping hacktivism', *Computer Fraud and Security*, 4, 1 April. Online. Available at: www.compseconline.com/publications/prodcfr.htm

Joseph, M. (23 December 2000) 'Both sides hacked over Kashmir', Wired News. Online. Available at: www.wired.com/news/politics/0,1283,40789,00.html.

Joseph, M. (3 June 2002) 'Tech king next Indian president?', Wired.com. Online. Available at: www.wired.com/news/politics/0,1283,52731,00.html

Kahn, R. and Kellner, D. (2004) 'New media and internet activism: from the "Battle of Seattle" to blogging', *New Media and Society*, 6(1), London and Thousand Oaks: Sage.

Kahney, L. (21 January 2003) 'Internet strokes anti-war movement', Wired.com. Online. Available at: www.wired.com/news/culture/0,1284,57310,00.html

Kahney, L. (17 March 2003) 'Citizen reporters make the news', Wired.com. Online. Available at: www.wired.com/news/culture/0,1284,58856,00.html

Kahney, L. (24 April 2004) 'Futurist fears end of innovation', Wired.com. Online. Available at: www.wired.com/news/politics/0,1283,58601,00.html

Kalathil, S. and Boas T.C. (July 2001) 'The Internet and state control in authoritarian regimes: China, Cuba, and the counterrevolution', Carnegie Endowment. Online. Available at: www.carnegieendowment.org/publications/index.cfm?fa=view&id=728 &prog=zgp

Karatzogianni, A. (2004a) 'The politics of cyberconflict', *Journal of Politics*, 24(1): 46–55.

Karatzogianni, A. (2004b) 'The effects of the internet on the Iraq war', *Cultural Technology and Policy* vol. 1, January 2004.

Karim, K.H. (2002) 'Diasporas and their communication networks: Exploring the broader context of transnational narrowcasting'. The Nautilus Institute: Virtual diasporas and global problem solving project papers. Online. Available at: www.nautilus.org/virtual-diasporas/paper/Karim.html.

Kattan, V. (29 March 2005) 'BBC reporting doesn't tell the whole story', electronicintifada.net. Online. Available at: http://electronicintifada.net/v2/article3724.shtml

Keane, J. (1996) *Reflections on Violence*, London and New York: Verso.

Kettmann, S. (10 January 2001) 'Deep thinking on the "Inter-Fada"', Wired.com. Online. Available at: www.wired.com/news/politics/0,1283,44919,00.html

Kettmann, S. (3 July 2001) 'Most hacking hides real threats', Wired.com. Online. Available at: www.wired.com/news/politics/0,1283,44955,00.html

Kettmann, S. (18 July 2001) 'Europe's spin on web reporting', Wired.com. Online. Available at: www.wired.com/news/culture/0,1284,45227,00.html

Kettmann, S. (13 July 2002) 'Make hate not anti-globalisation', Wired.com. Online. Available at: www.wired.com/news/politics/0,1283,53844,00.html

Khalizad, Z., White, J.P. and Marshall, W. (1999) *The Changing Role of Information Warfare*, California: Rand.

Khan, A. (1996) *The Extinction of Nation States: A World without Borders*, The Hague, London and Boston: Kluwer Law International.

Kidd, D. (2003) 'Indymedia.org.' in McCaughey, M. and Ayers, M. (eds), *Cyberactivism: Online Activism in Theory and Practice*, New York and London: Routledge.

King, B. (25 July 2002) 'Fear and lockdown in America', Wired.com. Online. Available at: www.wired.com/news/digiwood/0,1412,54099,00.html

Kiss, J. (14 October 2004) 'Student jailed for web pics', www.journalism.co.uk. Online. Available at: www.journalism.co.uk/news/story1099.shtml

Kiss, J. (18 March 2005) 'Bloggers rally to support voice of dissent in Iran', www.journalism.co.uk. Online. Available at: www.journalism.co.uk/news/story1375.shtml

Klandermans, B. (October 1984) 'Mobilisation and participation: Social-psychological expansions of resource mobilization theory', *American Sociological Review*, 49: 583–600.

Klandermans, B. (1994) 'Transient identities? Membership patterns in the Dutch peace movement' in Larana, E., Johnston, H. and Gussfield, J.R. (eds), *New Social Movements: From Ideology to Identity*, Philadelphia: Temple University Press.

Klandermans, B. and Goslinga, S. (1996) 'Media discourse, movement publicity, and the generation of collective action frames: Theoretical and empirical exercises in meaning construction' in McAdam, D., McCarthy, J. and Zald, M. (eds), *Comparative Perspectives on Social Movements: Political Opportunities, Mobilizing Structures, and Structural Framings*, Cambridge: Cambridge University Press.

Kneen, C. (28 February 2000) 'Battle in Seattle', interweb-tech.com. Online. Available at: www.interweb-tech.com/nsmnet/docs/voices_kneen.htm

Knight, W. 'Key Arab news station knocked offline', *New Scientist*. Online. Available at: www.newscientist.com/article.ns?id=dn3537

Knightley, P. (2000) *The First Casualty: The War Correspondent as Hero and Mythmaker from Crimea to Kosovo*, London: Prion.

Knightley, P. (14 June 2003) 'Turning the tanks on the reporters', *Guardian*. Online. Available at: http://observer.guardian.co.uk/iraq/story/0,12239,977702,00.html

Ko, Y. (3 October 2003) 'Cyberwars in China', *Far Eastern Economic Review*. Online. Available at: www.feer.com/hg76dkg75jg/0210_03_p066current.html

Koppel, T. (7 May 1995) 'Impact of television on US Foreign policy', Congressional Hearing. Online. Available at: www.cdi.org/adm/834/transcript.html

Koprowski, G. (6 October 2004) 'The Web: Iraqi blogs building free speech', United Press International. Online. Available at: www.washtimes.com/upi-breaking/20041005-050048-8337r.htm

Koprowski, G. (23 March 2005) 'The Web: The battle of the bloggers,' www.upi.com. Online. Available at: www.washtimes.com/upi-breaking/20050322-100604-1950r.htm

Korea Herald (26 May 2005) Korea: Internet cannot substitute democratic process'. Online. Available at: www.asiamedia.ucla.edu/tsunami/article.asp?parentid=24828

Kotadia, M. (11 January 2005) 'Security researcher to be jailed for finding bugs in software?' ZDNet Australia. Online. Available at: www.zdnet.com.au/news/security/0,2000061744,39176657,00.htm

Krebs, B. (20 March 2003) 'Websites vandalised with antiwar messages', *Washington Post*. Online. Available at: www.washingtonpost.com/wp-dyn/articles/A62865-2003Mar20.html

Kriesi, H. (March 1989) 'New social movements and the new class in the Netherlands', *American Journal of Sociology*, 94(5): 1078–1116.

Kroker, A. and Weinstein, M. (1994) *Data Trash: The Theory of Virtual Class*, New York: St. Martin's Press.

Kurtz, H (22 March 2003) 'Webloggers, signing on as war correspondents', *Washington Post*. Online. Available at: www.washingtonpost.com/ac2/wp-dyn/A47852-2003Jul25?language=printer

Kurtz, H. (13 August 2003) 'Dean defence forces: lobbing email at the enemy', *Washington Post*. Online. Available at: www.washingtonpost.com/ac2/wp-dyn/A47852-2003Jul25?language=printer

Kuznetsov, S. (30 October 2002) 'Russia: dial "H" for hostage', Wired.com. Online. Available at: www.wired.com/news/culture/0,1284,56073,00.html.

Laclau, E. (ed.) (1994) *The Making of Political Identities*, London and New York: Verso.

Laclau, E. and Mouffe, C. (1985) *Hegemony and Social Strategy: Towards a Radical Democratic Politics*, London and New York: Verso.

Landow, G.P. (1997) *Hypertext 2.0: The Convergence of Contemporary Critical Theory and Technology*, 2nd edn, Baltimore: Johns Hopkins University Press.

Langman, L., Morris, D. and Zalewski, J. (2002) 'Globalization, domination and cyberactivism', in Wilma A. Dunaway (ed.), *The 21st Century World-System: Systemic Crises and Antisystemic Resistance*, Westport, CT: Greenwood Press.

Lasker, J. (18 April 2005) 'US Military's elite hacker crew', Wired.com. Online. Available at: www.wired.com/news/privacy/0,1848,67223,00.html

Leadbeater, C. (4 December 2004) 'The internet's way forward', *Financial Times*. Online. Available at: http://specials.ft.com/creativebusiness/FT3XAQN4RUC.html

Lebert, J. (2003) 'Wiring human rights activism' in McCaughey, M. and Ayers, M. (eds), *Cyberactivism: Online Activism in Theory and Practice*, New York and London: Routledge.

Lebowitz, R. (22 May 2003) 'Can Internet technology still revolutionize activism?', Digital Freedom Network. Online. Available at: www.bobsonwong.com/dfn/workshop/elect-act.htm

Lee, J. (23 February 2003) 'How the protesters mobilized', *New York Times*. Online. Available at: http://homepages.wmich.edu/~jswanson/03s/100/articles/mar11art1.htm

Leech, G. (1 October 2005) 'Scared into silence', *Columbia Journal*. Online. Available at: www.colombiajournal.org/colombia200.htm

Leister, B. (4 August 2003) 'Net brings together campaign supporters', News Hour media unit. Online. Available at: http://press.meetup.com/archives/000312.html

Left, S. (4 May 2001) 'Chinese and American hackers declare "cyberwar"', Guardian unlimited. Online. Available at: www.guardian.co.uk/china/story/ 0,7369, 486089,00.html

Le Grignou, B. and Patou, C. (2004) 'ATTAC(k)ing expertise' in Van de Donk, W., Loader, B., Nixon, P.and Rucht, D. (eds), *Cyberprotest: New Media, Citizens and Social Movements*, London and New York: Routledge.

Leicester, J. (21 September 2004) 'Pentagon restricts overseas voters', Associated Press. Online. Available at: www.commondreams.org/headlines04/0921-03.htm

Lemos, R. (6 November 2000) 'Hacktivism: Mideast war heats up', zdnet.com. Online. Available at: http://news.zdnet.com/2100-9595_22-525308.html

Lemos, R. (11 November 2003) 'Attack on SCO's servers intensifies', news.com. Online. Available at: http://news.com.com/2100-7355_3-5120706.html

Lenz, M.W. (15 August 2003) 'Internet rejuvenates political movement', seacostonline.com. Online. Available at: www.seacoastonline.com/2003news/08062003/news/43449.htm

Leonard, A. (21 March 2002) 'Will the Net save China?', Salon.com. Online. Available at: www.salon.com/tech/books/2002/03/21/china_dawn/

Lesser, I., Hoffman, B., Arquilla, J., Ronfeldt, D., Zanini, M. and Jenkins, B.M. (1999) *Countering the New Terrorism*, California: Rand.

Lettice, J. (11 November 2004) 'We seize servers, you can't complain', The Register. Online. Available at: www.theregister.co.uk/2004/11/11/gov_indymedia_response/

Levine, R. (September 2003) 'Neal Stephenson rewrites history', Wired.com, Issue 11. Online. Available at: www.wired.com/wired/archive/11.09/history.html

Levy, P. (1995) 'Qu'est-ce que le Virtuel?', Editions La Decouverte, trans. Kritiki (1999) Athens.

Leyden, J. (17 April 2004) 'Middle East conflict spills over to cyberspace', The Register. Online. Available at: www.theregister.co.uk/2002/04/17/middle_east_conflict_spills_ over

Libicki, M. and Shapiro, J. (2001) *Strategic Appraisal: The Changing Role of Information Warfare*, California: Rand.

Lim, L. (2 July 2004) 'China to censor text messages'. Online. Available at: http://news.bbc.co.uk/2/hi/asia-pacific/3859403.stm

Ling, R. (1996) 'Cyber McCarthyism: Witch hunts in the living', *Electronic Journal of*

Sociology. Online. Available at: http://www.sociology.org/content/vol002.001/ling.html

Loader, B. (ed.) (1997) *The Governance of Cyberspace*, London and New York: Routledge.

Loader, B. (ed.) (1998) *Cyberspace Divide: Equality, Agency and Policy in the Information Society*, London: Routledge.

Luard, T. (30 January 2004) 'Chinese activists evade web controls', BBCnews online. Online. Available at: http://news.bbc.co.uk/1/hi/world/asia-pacific/3440911.stm

Lukes, S. (1970) *Power: A Radical View*, London and Basingstoke: Macmillan.

Manjoo, F. (7 June 2002) 'Do dots connect to police state?', Wired.com. Online. Available at: www.wired.com/news/politics/0,1283,53037,00.html

Mann, C. and Stewart, F. *Internet Communication and Qualitative Research: A Handbook for Researching Online*, London: Sage.

Mansbach, R. (2000) *The Global Puzzle*, 3rd edn, Boston: Houghton Mifflin.

Mansell, R. (2001) *Inside the Communication Revolution*, Oxford: Oxford University Press.

March, S. (28 February 2000) 'Battle in Seattle', interweb-tech.com. Online. Available at: www.interweb-tech.com/nsmnet/docs/voices.htm

March, S. (4 September 2005) 'Community organising on the Internet: Implications for social work practitioners', M.S.W. Online. Available at: www.interweb.com/nsmnet/docs/march

Margolis, M. and Resnick, D. (2000) *Politics As Usual*, Thousand Oaks, London and New York: Sage.

Mark, R. (29 April 2002) 'Internet voting improves participation', internet.com. Online. Available at: www.clickz.com/stats/sectors/software/article.php/1008791

Mattelart, A. and Mattelart, M. (1992) *Rethinking Media Theory*, Minneapolis: University of Minnesota Press.

Matthews, W. (25 July 2002) 'Al-Qaeda cyber alarm sounded', Wired.com. Online. Available at: www.crimeresearch.org/news/2002/07/Mess2601.htm

Mayfield, K. (14 October 2001) 'Chronicling attacks on the web', Wired.com. Online. Available at: www.wired.com/news/conflict/0,2100,47184,00.html

Mazzetti, M. 'PR meets psy-ops in war on terror', *Los Angeles Times*. Online. Available at: www.commondreams.org/headlines04/1201-01.htm

McAdam, D. (1996) 'The framing function of movement tactics: Strategic dramaturgy in the American civil rights movement' in McAdam, D., McCarthy, J.D. and Zald, M.N. (eds), *Comparative Perspectives on Social Movements: Political Opportunities, Mobilizing Structures, and Structural Framings*, Cambridge: Cambridge University Press.

McAdam, D., McCarthy, J. and Zald, M. (eds) (1996) *Comparative Perspectives on Social Movements: Political Opportunities, Mobilizing Structures, and Structural Framings*, Cambridge: Cambridge University Press.

McCarthy, J. (1996) 'Constraints and opportunities in adopting, adapting and inventing' in McAdam, D., McCarthy, J.D. and Zald, M.N. (eds), *Comparative Perspectives on Social Movements: Political Opportunities, Mobilizing Structures, and Structural Framings*, Cambridge: Cambridge University Press.

McCarthy, J. and Zald, M. (May 1977) 'Resource mobilization and social movements: A partial theory', *The American Journal of Sociology*, 82(6): 1212–1241.

McCarthy, J., McPhail, C. and Smith, J. (January 1996) 'Images of protest: Dimensions of selection bias in media coverage of the Washington demonstrations 1982 and 1991', *American Sociological Review*, 61(3): 478–499

McCarthy, K. (9 July 2003) 'This is the future of online newspapers'. Online. Available at: www.theregister.co.uk/2003/07/09/this_is_the_future/

McCaughey, M. and Ayers, M. (2003) *Cyberactvism: Online Activism in Theory and Practice*, New York and London: Routledge.

McClellan, J. (19 February 2003) 'War on the web', *Guardian*. Online. Available at: www.guardian.co.uk/online/story/0,3605,898661,00.html

McCullagh, D. (6 September 2001) 'Uncle Sam wants his geeks back', Wired.com. Online. Available at: www.wired.com/news/politics/0,1283,46569,00.html

McCullagh, D. and Zarate R. (27 February 2002) 'Hack a PC, get life in jail', Wired.com. Online. Available at: www.wired.com/news/politics/0,1283,50708,00.html

McGrath, D. (13 July 2002) 'Tunisian net dissident jailed', Associated Press. Online. Available at: www.wired.com/news/politics/0,1283,53186,00.html

McGregor, B. (1997) *Live, Direct and Biased? Making Television News in the Satellite Age*, London and New York: Arnold.

McKenna and Bargh, J. (1998) 'Coming out in the age of the Internet: Identity, demarginalization through virtual group participation', *Journal of Personality and Social Psychology*, 75: 681–694.

McLaughlin, K. (9 September 2004) 'Internet prods Asia to open up', *Christian Science Monitor*. Online. Available at: www.csmonitor.com/2004/0909/p06s01-woap.html

McMillan, R. (27 March 2003) 'Wartime internet security is "business as usual"', *Washington Post*. Online. Available at: www.crime-research.org/news/2003/03/ Mess2802.html

McNay, L. (1994) *Foucault: A Critical Introduction*, Cambridge: Polity Press.

McWilliams, B. (29 July 2002) 'Fluffy bunny no longer energized', Wired.com. Online. Available at: www.wired.com/news/technology/0,1282,54040,00.html

McWilliams, B. (15 November 2002) 'Dot-Mil hacker's download mistake', Wired.com. Online. Available at: www.wired.com/news/technology/0,1282,56392,00.html

McWilliams, B. (5 December 2002) 'Messages to al-Qaeda called fake', Wired.com. Online. Available at: www.wired.com/news/culture/0,1284,56715,00.html

McWilliams, B. (20 March 2003) 'Leaked bug alerts cause a stir', Wired.com. Online. Available at: www.wired.com/news/infostructure/0,1377,58106,00.html

McWilliams, B. (2 June 2003) 'North-Korea's school for hackers', Wired.com. Online. Available at: www.wired.com/news/politics/0,1283,59043,00.html

McWilliams, B. (30 July 2003) 'Iraqis log on to voice chat', Wired.com. Online. Available at: www.wired.com/news/technology/0,1282,59786,00.html

Meacher, M. (6 September 2003) 'The 9/11 attacks gave the US an ideal pretext to use force to secure its global domination'. Online. Available at: http://politics. guardian.co.uk/iraq/comment/0,12956,1036687,00.html

Meek, C. (16 December 2004) 'Milestone in online journalism', journalism.co.uk. Online. Available at: www.journalism.co.uk/features/story1183.shtml

Melucci, A. (1989) *Nomads of the Present: Social Movements and Individual Needs in Contemporary Society*, ed. John Keane and Paul Mier, Philadelphia: Temple University Press.

Mercer, D., Mungham, G. and Williams, K. (1987) *The Fog of War: The Media on the Battlefield*, London: Heinnemann.

Mermin, J. (Autumn 1997) 'Television news and American intervention in Somalia: The myth of a media driven foreign policy', *Political Science Quarterly*, 112(3): 385–403.

Messmer, E. (5 April 1999) 'Serb supporters Sock it to NATO and US computers', Network World. Online. Available at: www.networkworld.com/news/1999/0405nato. html

Meyerson, H. (28 Feburary 2000) 'The battle of Seattle', *LA Weekly*/interweb-tech.com. Online. Available at: www.laweekly.com/ink/00/02/powerlines-meyerson.shtml

Meyrowitz, J. *No Sense of Place: The Impact of Electronic Media on Social Behaviour*, New York and Oxford: Oxford University Press.

Miami Herald (3 April 2001) 'Terrorists using the net to raise funds reach public'. Online.

Michaelides G. (2006). *Dynamics of Knowledge Sharing in the KDE Open Source Community*, unpublished PhD thesis, University of Nottingham.

Milone, M.G. (November 2002) 'Hactivism: Securing the national infrastructure', *The Business Lawyer*, 58(1): 383–415.

Minear, L., Scott, C. and Weiss, T. (1994) *The News Media, Civil War and Humanitarian Action*, Boulder and London: Lynne Rienner.

Minkoff, D. (October 1997) 'The sequencing of social movements', *American Sociological Review*, 62(5): 779–799.

Mintz, J. (6 December 2002) 'Al-Qaeda web site calls Israel new target', *Washington Post*. Online. Available at: http://memri.org/bin/media.cgi?ID=37102

Mollov, B., Schwartz, Steinberg, G. and Lavie, C. (undated) 'The impact of Israeli–Palestinian intercultural dialogue: Virtual and face to face'. Online. Available at: www.mevic.org/papers/mollovvirtual.html

Mooney, P. (23 April 2004) 'China's big mamas vs. online dissidents', *International Herald Tribune*. Online. Available at: www.iht.com/articles/2004/04/23/edmooney_ed3.php

Moore, P., Bormann, N., Charnock, G., Cozette, M., Elias, J., Hague, S., Hancock, J., Jackson, R., Mawdsley, J., Parmar, I., Wilkinson R. and Young, R. (2006) 'Normalising Empire, Ignoring Imperialism', Centre for International Politics, University of Manchester, working paper. Online. Available at: www.socialsciences.manchester.ac.uk/politics/research/research_groups/cip/cip_publications.htm

Moore, R.K. (20 October 2004) 'We the people, IndyMedia, and the neoliberal project', globalresearch.ca. Online. Available at: http://globalresearch.ca/articles/MOO410A.html

Morse, M. (1998) *Virtualities: Television, Media Art and Cyberculture*, Bloomington and Indianapolis Indiana University Press.

Mosaad, Mohamed (undated) 'The web community: A social movement in the cyberspace?', Department of Sociology, American University of Cairo. Online. Available at: www.mevic.org/papers/social-movements-cyberspace.html

Mosco, V. and Schiller, D. (2001) *Continental Order? Integrating North America for Cybercapitalism*, Lanham: Rowman and Littlefield.

Moss, J. (ed.) (1998) *The Later Foucault*, London: Sage.

Mouffe, C. (1993) *The Return of the Political*, London and New York: Verso.

Mouffe, C. (2000) *The Democratic Paradox*, London and New York: Verso.

Mueller, C. (October 1997) 'International press coverage of East German protest events', *American Sociological Review*, 62(5): 820–852.

Mulveron, J. and Yang, R. (1999) *The People's Liberation Army in the Information Age*, California: Rand.

Munson, C. (30 November 2004) '*Seattle Weekly* trashes anti-globalization movement', counterpunch.org. Online. Available at: www.counterpunch.org/munson 11302004.html

Murray, T. (2004) 'New media art and rhizomatic in-securities', Ctheory.net. Online. Available at www.ctheory.net/articles.aspx?id=420

Mustafa, W. (4 April 2003) 'Pakistan's netizens outsmart censors', Wired.com. Online. Available at: www.oneworld.net/article/archive/1866/340

Myers, M. (2001) 'Collective identity.org: Collective identity in online and offline feminist activist groups', thesis, Virginia Polytechnic Institute.

Negroponte, N. (1995) *Being Digital*, London: Hodder and Stoughton.

Negroponte, N. (24 September 2002) 'Being wireless,' Wired.com. Online. Available at: www.wired.com/wired/archive/10.10/wireless.html

Netto, A. 'Malaysia backtracks on its bloggers', *Asia Times*. Online. Available at: http://atimes.com/atimes/Southeast_Asia/FJ09Ae05.html

New Scientist (24 November 2004) 'The "blog" revolution sweeps across China', Online. Available at: www.newscientist.com/article.ns?id=dn6707

News.muzi.com (11 May 2005) 'Tokyo tightens cyber defense after protest'. Online. Available at: http://news.muzi.net/ll/english/1362332.shtml

Newsbycountry.asp (15 April 2003) 'UK web surfers prefer BBC, *Guardian*', newsbycountry.asp. Online. Available at: www.allbusiness.com/periodicals/article/552662-1.html

Newsbytes (17 June 2002) 'American companies fail to report intrusions'. Online. Available at: www.nua.com/surveys/index.cgi?f=VS&art_id=905357825&rel=true

Nip, J. (2004) 'The Queer Sisters and its electronic bulletin board' in Van de Donk, W., Loader, B., Nixon, P. and Rucht, D. (eds), *Cyberprotest: New Media, Citizens and Social Movements*, London and New York: Routledge.

Nua Analysis. (24 June 2002) 'E-government gap widens', nua.ie. Online. Available at: www.nua.com/surveys/analysis/weekly_editorial/archives/issue1no302.html

Nugent, D. (1995) 'Nortern Intellectuals and EZLN', *Monthly Review*, 47(3), July–August.

Nunes, M. (1995) *Baudrillard in Cyberspace: Internet, Virtuality, and Postmodernity, Style*, 29: 314–327.

Nuttal, C. (5 October 1998) 'India opens virtual front in Kashmir', BBConline. Online. Available at: http://news.bbc.co.uk/hi/english/sci/tech/newsid_186000/186952.stm

Nuttal, C. (16 April 1999) 'Kosovo – the Internet war', BBCnews. Online. Available at: http://news.bbc.co.uk/hi/english/world/from_our_own_correspondent/newsid_319000/319003.stm

Oakes, C. (15 April 2000) 'Balkan war in domain attacks?', Wired.com. Online. Available at: www.wired.com/news/politics/0,1283,35674,00.html

Offe, C. (1985) 'New social movements: Challenging the boundaries of institutional politics', *Sociological Research*, 52: 817–868.

O'Hara, K. (2002) 'The internet: a tool for democratic pluralism?', *Science as Culture*, 11(2): 287–298.

Olofsson, G. (1988) 'After the working class movement? An essay on what's "new" and what's "social" in the new social movements', *Acta*, 31: 15–34.

OpenNet Initiative (30 August 2004) 'Google search and cache filtering behind China's great firewall', opennetinitiative.net. Online. Available at: www.opennetinitiative.net/blog/?p=60

Ortner, E. (8 September 2003) 'Worm suspect says case against him exaggerated', msnbc.com. Online. Available at: www.msnbc.com/news/960926.asp

Ouyang Fei. (November 2004) 'Tapping into the TV system to broadcast truth clarification videos is a form of resistance against state-run terrorism', clearwisdom.net. Online. Available at: http://clearwisdom.net/emh/articles/2004/11/28/55033.html

Parkin, f. (1968) *Middle Class Realism*, Manchester: Manchester University Press.

Pateman, C. (1970) *Participation and Democratic Theory*, Cambridge: Cambridge University Press.

Penenberg, A. (1 September 2004) 'Site tracks political zeitgeist', Wired.com. Online. Available at: www.wired.com/news/culture/0,1284,64791,00.html

Phrack Magazine (December 1998) 'Pentagon blocks DoS attack'. Online. Available at: http://66.249.93.104/search?q=cache:k0bIosnh57wJ:www.phrack.org/

Pini, B., Brown, K. and Previte, J. (2004) 'Politics and identity in cyberspace: a case study of Australian women in agriculture online', in van de Donk, W., Loader, B.D., Nixon, P.G. and Richt, D. (eds) *Cyberprotest: New Media; Citizens and Social Movements*, London: Routledge.

Politics Online (16 May 2002) 'The birth of a nation: online', politicsonline.com. Online.

Politics Online (27 June 2002) 'The European Union student vote', politicsonline.com. Online. Available at: www.election-europe.com.

Politics Online (26 July 2003) The Moveon Primary: What it means, politicsonline.com. Online.

Posner, G. (2003) *Why America Slept*, New York: Random House.

Poster, M. (1995) 'Cyberdemocracy: Internet and the public sphere', Irvine: University of California. Online. Available at: www.humanities.uci.edu/mposter/writings/democ.html

Poster, M. (1995) *The Second Media Age*, Cambridge: Polity Press.

Poulsen, K. (11 July 2005) 'Feds fear air broadband terror', Wired.com. Online. Available at: www.wired.com/news/technology/0,1282,68147,00.html

Prosser, B.T. and Ward, A. (24 May 2002) 'Kierkegaard's "mystery of unrighteousness" in the information age', abdn.ac.uk. Online. Available at: www.abdn.ac.uk/philosophy/endsandmeans/vol5no2/prosser_ward.shtml

Pruitt, S. (9 October 2003) 'EU approves creation of cybersecurity agency', IDG News Service, London Bureau. Online. Available at: www.infoworld.com/article/03/10/09 HNeuagency_1.html

Raab, C.D. (1997) in Loader, B. (ed.) *The Governance of Cybercspace*, London: Routledge.

Ragusa, R. (10 June 2002) 'Text messaging: let the US mobile marketing phenomenon begin'. Online. Available at: www.bayoubuzz.com/political/articles/messaging_rand.htm

Rainie, L., Fox, S. and Fallowes, D. (undated) 'The Internet and the Iraq war: How online Americans have used the Internet to learn war news, understand events and promote their views', Pew Internet and American Life Project. Online. Available at: www.pewinternet.org/pdfs/PIP_Iraq_War_Report.pdf

Ramachandran, S. (31 May 2005) 'Jihad's latest "rag" hits the internet', *Asia Times*. Online. Available at: www.atimes.com/atimes/Middle_East/GC31Ak03.html

Raphael, C. (10 February 2005) 'Spinning media for government', corpwatch.org. Online. Available at: www.corpwatch.org/article.php?id=11836

Ratan, S. (17 August 2003) 'It's a flawed world after all', Wired.com. Online. Available at: www.wired.com/news/culture/0,1284,60063,00.html

Rathmell, A. (July 1997) 'Information warfare: the coming threat', The Newsletter for Criminal Analysts within NCIS, 2: 1. Online. Available at: www.kcl.ac.uk/orgs/icsa/Old/ncis.html

Rathmell, A. (2000) 'Information warfare and sub-state actors: an organizational approach', in Thomas, D. and Loader, B. (eds) *Cybercrime: Law Enforcement and Surveillance in the Information Age*, London: Routledge.

Raymond, E.S. (2001) *The Cathedral and the Bazaar: Musings on Linux and Open Source from an Accidental Revolutionary*, Sebastapol, CA: O'Reilly and Associates.

Rendall, S. and Broughel, T. (May/June 2003) 'Amplifying officials, squelching dissent: FAIR study finds democracy poorly served by war coverage', *Extra!* Online. Available at: www.fair.org/index.php?page=1145

Rengi Neer (14 November 2004) 'Symantec helps China crush freedom of information', indymedia.org. Online. Available at: http://newswire.indymedia.org/en/newswire/archive98.shtml

Reporters Without Borders (13 August 2004) 'President Gayoom cuts off Internet links with outside world'. Online. Available at: www.rsf.org/print.php3?id_article=11137

Reuters (24 April 2000) 'AOL founder: censor the net? Ha!', Wired.com. Online. Available at: www.wired.com/news/politics/0,1283,35854,00.html

Reuters (18 June 2002) 'China shuts down internet cafes', Wired.com. Online. Available at: www.wired.com/news/politics/0,1283,56993,00.html

Reuters (21 June 2002) 'Kremlin's new website stands up to hacker threats'. Online. Available at: www.linuxtoday.com/news_story.php3?ltsn=2002-06-22-007-26-NW-RH-SV

Reuters (9 September 2002) 'No Google for Chinese surfers', Wired.com. Online. Available at: www.wired.com/news/politics/0,1283,55030,00.html

Reuters (30 October 2003) 'UK plans to extradite spammers'. Online. Available at: www.wired.com/news/technology/0,1282,61021,00.html

Reuters (31 October 2003) 'China locks up net dissident'. Online. Available at: www.wired.com/news/culture/0,1284,61046,00.html

Reuters (19 February 2003) 'US plans for online defence'. Online. Available at: www.wired.com/news/technology/0,1282,57695,00.html

Reuters (6 March 2003) 'Media plot online war coverage'. Online. Available at: www.wired.com/news/conflict/0,2100,57933,00.html

Reuters (18 March 2003) 'Not just your average loser'. Online. Available at: www.wired.com/news/technology/0,1282,58102,00.html

Reuters (28 March 2003) 'War attacks tit for tat'. Online. Available at: www.wired.com/news/conflict/0,2100,58275,00.html

Reuters (16 June 2003) 'Britain's Labour party website hit by hackers'. Online. Available at: http://news.bbc.co.uk/1/hi/uk_politics/2993550.stm

Reuters (12 August 2003) 'New worm mocks "Billy" Gates', Wired.com. Online. Available at: www.wired.com/news/technology/0,1282,59987,00.html

Reuters (26 August 2003) 'Amazon fights brand "spoofing"', Wired.com. Online. Available at: www.wired.com/news/business/0,1367,60191,00.html

Reuters (12 September 2003) 'Hacker must live with parents', Wired.com. Online. Available at: www.wired.com/news/technology/0,1282,60429,00.html

Reuters (8 May 2004) 'Sasser worm suspect confesses'. Online. Available at: www.wired.com/news/infostructure/0,1377,63393,00.html

Reuters (18 November 2004) 'Iraq journo "held" in Fallujah'. Online. Available at: www.williambowles.info/ich/ich-1104-3.html

Reuters (23 March 2005) 'Nepal scribes evade censors with blogs'. Online. Available at: www.telegraphindia.com/1050324/asp/foreign/story_4531440.asp

Rheingold, H. (1991) *Virtual Reality*, London: Secker and Warburg.

Rheingold, H. (1994) *The Virtual Community: Surfing the Internet*, London: Minerva.

Rheingold, H. (2002) *Smart Mobs: The Next Social Revolution*, Perseus Publishing.

Richardo, N. (1997) 'New social movements: a critical review', *Annual Review of Sociology*, 23: 411–430.

Ronfeldt, D., Arquilla J., Fuller, G.E. and Fuller, M. (1998) *The Zapatista Social Netwar in Mexico*, California: Rand.

Rothstein, R.L. (1999) *After the Peace*, Boulder and London: Lynne Rienner.

Rozen, L. (7 August 2003) 'Forums point the way to jihad', Wired.com. Online. Available at: www.wired.com/news/culture/0,1284,59897,00.html

Rosenkrands, J. (2004) 'Politicizing *Homo economicus*' in Van de Donk, W., Loader, B., Nixon, P. and Rucht, D. (eds), *Cyberprotest: New Media, Citizens and Social Movements*, London and New York: Routledge.

Rucht, D. (1996) 'The impact of national contexts on social movement structures: A cross-movement and cross-national comparison' in McAdam, D., McCarthy, J. and Zald, M. (eds), *Comparative Perspectives on Social Movements: Political Opportunities, Mobilizing Structures, and Structural Framings*, Cambridge: Cambridge University Press.

Rucht, D. (2004) 'The quadruple "A": Media strategies of protest movements since the 1960s' in Van de Donk, W., Loader, B., Nixon, P. and Rucht, D. (eds), *Cyberprotest: New Media, Citizens and Social Movements*, London and New York: Routledge.

Salecl, R. (1996) 'National identity and social moral Majority', in Eley, G. and Suny (eds), *Becoming National: A Reader*, Oxford: Oxford University Press.

Salter, L. (2003): 'Democracy, social movements, and the Internet' in McCaughey, M. and Ayers, M. (eds), *Cyberactivism: Online Activism in Theory and Practice*, New York and London: Routledge.

Sandhana, L. (30 July 2003) 'I think, therefore I communicate', Wired.com. Online. Available at: www.wired.com/news/medtech/0,1286,59737,00.html

Savetibet.org (28 October 2003) 'Chinese internet group found spying on Tibetan government computers'. Online. Available at: www.savetibet.org/news/newsitem.php?id=533

Scahill, J. (9 September 2004) 'Indymedia and the text message jihad', counterpunch.org. Online. Available at: www.counterpunch.org/scahill09092004.html

Schechter, D. (25 March 2005) 'Miscovering anti-war protest', zmag.org. Online. Available at: www.zmag.org/sustainers/content/2005-03/25schechter.cfm

Scheeres, J. (4 March 2002) 'Columbia's cyber (un)civil war', Wired.com. Online. Available at: www.wired.com/news/politics/0,1283,50748,00.html

Scheeres, J. (25 March 2002) 'Cuba bans PC sales to public', Wired.com. Online. Available at: www.wired.com/news/politics/0,1283,51270,00.html

Scheeres, J. (19 June 2002) 'Lech Walesa: tech freedom fighter', Wired.com. Online. Available at: www.wired.com/news/politics/0,1283,53299,00.html

Scheeres, J. (2 August 2002) 'Cautious Kabul dabbles with net', Wired.com. Online. Available at: www.wired.com/news/culture/0,1284,54285,00.html

Scheeres, J. (9 November 2002) 'Europeans outlaw net hate speech', Wired.com. Online. Available at: www.wired.com/news/business/0,1367,56294,00.htm

Scheeres, J. (25 March 2003) 'Blair tagged as privacy threat', Wired.com. Online. Available at: www.wired.com/news/politics/0,1283,58189,00.html

Scheeres, J. (1 August 2003) 'Spanish firms target file traders', Wired.com. Online. Available at: www.wired.com/news/digiwood/0,1412,59720,00.html

Scheeres, J. (24 February 2004) 'Net dissidents jailed in China'. Online. Available at: www.wired.com/news/politics/0,1283,62391,00.html

Scheeres, J. (22 June 2004) 'Progress report for net censors', Wired.com. Online. Available at: www.wired.com/news/politics/0,1283,63940,00.html

Schmelzer, P. (12 August 2004) 'Anarchy at the RNC', infoshop.org. Online. Available at: www.alternet.org/rights/19541/

Scott, A. (1990) *Ideology and the New Social Movements*, Boston: Unwin Hyman.

Shachtman, N. (18 July 2002) 'Why countries make sites unseen', Wired.com. Online. Available at: www.wired.com/news/politics/0,1283,53933,00.html

Shachtman, N. (30 August 2002) 'Hackers being jobbed out of work', Wired.com. Online. Available at: www.wired.com/news/culture/0,1284,54838,00.html

Shachtman, N. (14 November 2002) 'Study makes less of hack threat', Wired.com. Online. Available at: www.wired.com/news/politics/0,1283,59818,00.html

Shachtman, N. (4 December 2002) 'An inside look at China filters', Wired.com. Online. Available at: www.wired.com/news/politics/0,1283,56699,00.html

Shachtman, N. (20 December 2002) 'Terrorists on the net? Who cares?', Wired.com. Online. Available at: www.wired.com/news/infostructure/0,1377,56935,00.html

Shachtman, N. (09 April 2003) 'Urban combat takes street smarts', Wired.com. Online. Available at: www.wired.com/news/conflict/0,2100,58381,00.html

Shachtman, N. (30 July 2003) 'The case for terrorism futures', Wired.com. Online. Available at: www.wired.com/news/politics/0,1283,59818,00.html

Shachtman, N. (13 September 2003) 'Sex sites sick of getting screwed', Wired.com. Online. Available at: www.wired.com/news/business/0,1367,60423,00.html

Shachtman, N. (16 October 2003) 'Spies attack White House secrecy', Wired.com. Online. Available at: www.wired.com/news/business/0,1367,60836,00.html

Shachtman, N. (17 August 2004) 'Hackers take aim at GOP', Wired.com. Online. Available at: www.wired.com/news/politics/0,1283,64602,00.html

Shachtman, N. (21 September 2004) 'Hack attack gums up Authorize.Net', Wired.com. Online. Available at: www.wired.com/news/infostructure/0,1377,65039,00.html

Shankar Gupta (10 March 2005) 'Bloggers protest possible FEC crackdown', mediapost.com. Online. Available at: http://publications.mediapost.com/index.cfm?fuseaction=Articles.showArticleHomePage&art_aid=28049

Shaw, M. (1996) *Civil Society and Media in Global Crises: Representing Distant Violence*, London and New York: Pinter.

Sherriff, L. (1 November 2004) 'China shuts 1,600 cybercafes', The Register. Online. Available at: www.theregister.co.uk/2004/11/01/china_net_crackdown/

Shils, E. (1975) *Center and Periphery*, Chigago: The University of Chicago Press.

Shimeall, T., Dunlevy, C. and Williams, P. (May 2001) *Intelligence Analysis for Internet Security: Ideas, Barriers and Possibilities*, Certo Analysis Center, Carnegie Mellon University.

Silverman, J. (6 February 2004) 'Iran's most wanted: filmmakers', Wired.com. Online. Available at: www.wired.com/news/digiwood/0,62179-1.html

Simon, E. (26 September 2004) 'Soldiers' war blogs detail life in Iraq', Associated Press. Online. Available at: www.fortwayne.com/mld/newssentinel/9761439.htm

Simons, J. (1995) *Foucault and the Political*, London and New York: Routledge.

Singel, R. (24 November 2003) 'Congress expands FBI spying power', Wired.com. Online. Available at: www.wired.com/news/politics/0,1283,61341,00.html

Singel, R. (2 February 2004) 'Net politics down but not out', Wired.com. Online. Available at: www.wired.com/news/politics/0,1283,62123,00.html

Singer, M. (30 November 2004) 'China's Google block sparks media group's protest'. Online. Available at: www.internetnews.com/ent-news/article.php/3441781

Sipress, A. (14 December 2004) 'An Indonesian's prison memoir takes war into cyberspace', *Washington Post* Foreign Service. Online. Available at: www.washingtonpost.com/wp-dyn/articles/A62095-2004Dec13.html

Slevin, J. (2000) *The Internet and Society*, Cambridge: Polity Press.

Smith, M. and Kollock, P. (1999) *Communities in Cyberspace*, London and New York: Routledge.

Snow, D., Zurcher, L. and Olson, S.E. (October 1980) 'Social networks and social movements: A microstructural approach to differential recruitment', *American Sociological Review*, 45(5): 787–801.

Snyder, D. and Kelly, W. (February 1977) 'Conflict intensity, media sensitivity and the validity of newspaper data', *American Sociological Review*, 42(1): 105–123.

Sola Pool de I (1990) *Technologies without Boundaries*, Cambridge: Harvard University Press.

Sophos (15 September 2003) 'Two British men charged in Trojan horse case, Sophos anti-virus comments', sophos.com. Online. Available at: www.sophos.com/pressoffice/news/articles/2003/09/va_threatkrew.html

Starr, B. and Shaughnessy, L. (5 February 2005) 'Pentagon sites: Journalism or propaganda?', CNN. Online. Available at: www.cnn.com/2005/ALLPOLITICS/02/04/web.us/

Statesman, The (21 August 2001) 'Pro-Pakistan hackers deface centre's venture capital site'. Online. Available at: www.srijith.net/indiacracked/media/statesman_24082001.pdf

Sterling, B. (1992) *The Hacking Crackdown: Law and Disorder on the Electronic Frontier*, London: Viking.

Stillman, L. (October 1996) 'Citizens of the world or citizens of a community: just where is the Internet heading? Online. Available at: http://webstylus.net/?q=node/40

Strupp, J. (19 March 2003) 'Study: Media self-censored some Iraq coverage', editorandpublisher.com. Online. Available at: www.camerairaq.com/2005/03/study_media_sel.html

Strupp, J. (7 January 2005) 'Five embeds booted out of Iraq in recent months', editorpublisher.com. Online. Available at: www.editorandpublisher.com/eandp/article_brief/eandp/1/1000748266

Sturgeon, W. (16 June 2003) 'Italian police shut down piracy ring', CNET news.com. Online. Available at: http://news.com.com/Italian+police+shut+down+piracy+ring/2100-1012_3-1017776.html

Sutherland, J. (7 October 2002) 'Campus watch', Wired.com. Online. Available at: www.guardian.co.uk/Columnists/Column/0,5673,806021,00.html

Swartz, J. (18 March 2003) 'Iraq war could herald a new age of web-based news coverage', *USA Today*. Online. Available at: www.usatoday.com/tech/news/2003-03-18-iraq-internet_x.htm

Swett, C. (July 1995) 'Strategic assessment: The internet'. Online. Available at: www.fas.org/cp/swett.html

Taggart, S. (8 March 2002) 'Irish eyes smile on dot-TP', Wired.com. Online. Available at: www.wired.com/news/politics/0,1283,50659,00.html

Taira, C. (14 October 2004) '10 big stories the mainstream media missed', lasvegascitylife.com. Online. Available at: www.alternet.org/columnists/story/25824/

Talbot, D. (February 2005) 'Terror's server', technologyreview.com. Online. Available at: www.technologyreview.com/articles/05/02/issue/feature_terror.asp

Tarrow, S. (1983) 'Struggling to reform: Social movements and policy change during cycles of protest', *West. Soc. Pap.*, no. 15, Ithaca, NY: Cornell University Press.

Tarrow, S. (1988) 'National politics and collective action', *Annual Review of Sociology*, 14: 421–440.

Tarrow, S. (March 1991) 'Aiming at a moving target: social science and the recent rebellion in Eastern Europe', *Political Science and Politics*, 24(1): 12–20.

Tarrow, S. (1996) 'States and opportunities: The political structuring of social movements' in McAdam, D., McCarthy, J. and Zald, M. (eds), *Comparative Perspectives on*

Social Movements: Political Opportunities, Mobilizing Structures, and Structural Framings, Cambridge: Cambridge University Press.

Tarrow, S. (1998) *Power in Movement: Social Movements and Contentious Politics*, Cambridge: Cambridge University Press.

Taylor, Paul (1999) *Hackers*, London and New York: Routledge.

Taylor, Paul (3 October 2002) 'The pornographic barbarism of the self-reflecting sign'. Online. Available at: www.londonconsortium.com/events/files/taylortalk.doc

Taylor, Paul and Jordan, T. (2004) *Hacktivism and Cyberwars: Rebels with a Cause?*, London: Routledge.

Taylor, Phil (1992) *War and the Media Propaganda and Persuasion in the Gulf War*, Manchester and New York: Manchester University Press.

Taylor, V. and Whittier, N. (1992) 'Collective identity and lesbian feminist mobilization', in A. Morris and C. Mueller (eds), *Frontiers of Social Movement Theory*, New Haven: Yale University Press.

TechNews.com (3 October 2002) 'Worldwide cyber-security breaches on the up', nua.ie/surveys. Online. Available at: www.nua.com/surveys/index.cgi?f=VS& art_id=905358144&rel=true

Terdiman, D. (23 April 2004) 'Dems hold the high ground online', Wired.com. Online. Available at: http://wired.com/news/politics/0,1283,63183,00.html

Terdiman, D. (12 August 2004) 'Text messages for critical masses', Wired.com. Online. Available at: www.wired.com/news/politics/0,1283,64536,00.html

Theodoulou, M. (23 June 2003) 'Proliferating Iranian weblogs give voice to taboo topics', *Christian Science Monitor*. Online. Available at: www.csmonitor.com/2003/ 0623/p07s02-wome.html

Thiel, S. (September 1998) 'The online newspaper: A postmodern medium', *Journal of Electronic Publishing*, 4(1). Online. Available at: www.press.umich.edu/jep/04-01/thiel.html

Thomas, D. and Loader, B. (eds) (2000) *Cybercrime: Law Enforcement, Security and Surveillance in the Information Age*, London and New York: Routledge.

Thomas, L. (Summer 2004) 'Electronic resistance: Young Iranians go digital in their quest for freedom', Smygo list. Online. Available at: www.towardfreedom.com/home/ content/view/49/72/

Thompson, M. (1994) 'Forging war: The media in Serbia, Croatia and Bosnia-Herzegovina: Article 19', International Centre Against Censorship, London: The Bath Press.

Thompson, N. (18 June 2002) 'Machined politics', washingtonmonthly.com. Online. Available at: www.washingtonmonthly.com/features/2001/0205.thompson.html

Tidwell, A.C. (1998) *Conflict Resolved?* London and New York: Pinter.

Tilly, C. (1978) *From Mobilization to Revolution*, Reading, MA: Addison-Wesley.

Tilly, C. (1984) 'Social movements and national politics' in C. Bright and S. Harding (eds), *State-making and Social Movements*, Ann Arbor: Michigan University Press.

Toffler, A. (1980) *The Third Wave*, New York: Morrow.

Tomlin, J. (16 May 2003) 'Censored in the desert', Press Gazette.

Torfing, J. (1999) *New Theories of Discourse: Laclau, Mouffe, Žižek*, Oxford: Blackwell.

Tormey, S. (2004) *Anti-capitalism: A Beginner's Guide*, Oxford: Oneworld.

Touraine, A. (1985) 'An introduction to the study of social movements', *Sociological Research*, 52: 749–788.

TrendLabs Global Antivirus and Research Centre (12 September 2003) 'Trend Micro weekly virus report' trendmicro.com. Online.

Trotochaud, M. and McDowell, R. (22 January 2005) 'The invasion of Falluja: A study in the subversion of truth', commondreams.org. Online. Available at: www.common-dreams.org/views05/0122-09.htm

Trotsky, L. 'The new course'. Online. Available at: www.marxists.org/archive/trotsky/1923-nc/index.htm

Tsagarousianou, R., Tambini, D. and Bryan, C. (eds) (1998) *Cyberdemocracy: Technology, Cities and Civic Networks*, London and New York: Routledge.

Tsuruoka, D. (17 March 2003) 'Internet, Wireless to play key role in an Iraq war', yahoonews. Online. Available at: www.nlectc.org/justnetnews/03202003.html

Turner, R. and Killian, L. (1972) *Collective Behavior*, Englewood Cliffs, NJ: Prentice Hall.

Twist, J. (8 July 2005) 'Mobiles capture blast aftermath', BBCnews. Online. Available at: http://news.bbc.co.uk/2/hi/technology/4663561.stm

USA Today (15 August 2003) 'Groups campaign online against Burmese dictators'. Online. Available at: www.usatoday.com/tech/world/2003-07-11-mission-to-burma_x.htm

USA Today (August 2003) 'Efforts to put government online could backfire'. Online.

USA Today (7 May 2002) 'US forces using Net to contact home', Wired.com. Online.

Van de Donk, W., Loader, B., Nixon, P. and Rucht, D. (2004) *Cyberprotest: New Media, Citizens and Social Movements*, London and New York: Routledge.

Van Dijk, T. (1997) *Discourse as Social Interaction, Discourse Studies: A Multi-disciplinary Introduction*, Vol. 2, London: Sage.

Van Aelst, P. and Walgrave, S. (2004) 'New media, new movements?' in Van de Donk, W., Loader, B., Nixon, P. and Rucht, D. (eds), *Cyberprotest: New Media, Citizens and Social Movements*, London and New York: Routledge.

Van Handel, J.L. (28 February 2000) 'Battle in Seattle', interweb-tech.com. Online. Available at: www.interweb-tech.com/nsmnet/docs/voices.htm

Varghese, S. (27 July 2005) 'ISP "censored" antiwar email', *Sydney Morning Herald*.

Vatis, M. (2001) 'Cyberattacks during the war on terrorism: A predictive analysis', Dartmouth: Institute for Security Technology Studies, Dartmouth College.

Vegh, S. (2003) 'Classifying forms of online activism' in McCaughey, M. and Ayers, M. (eds), *Cyberactivism: Online Activism in Theory and Practice*, New York and London: Routledge.

Venzke, B. (4 August 1996) 'Information warrior', Wired.com. Online. Available at: www.wired.com/wired/archive/4.08/schwartau.html

Verton, D. (3 May 1999) 'New cyberterror threatens AF', Federal Computer Week. Online. Available at: www.fcw.com/fcw/articles/1999/FCW_050399_431.asp

Villafania, A. (7 January 2005) 'Internet activists to hold virtual strike at Gelmart website', www.inq7.net. Online. Available at: http://news.inq7.net/infotech/index.php?index=1&story_id=23498

Wagner, J. (5 December 2002) 'Development news: airline database posted on defacement', internetnews.com. Online. Available at: www.internetnews.com/dev-news/article.php/1013341

Wagner, J. (5 February 2002) 'Development news: navy brass latest hacking victim', internetnews.com. Online. Available at: www.internetnews.com/dev-news/article.php/1120971

Wagner, M. (3 October 2002) 'Fighting net censorship abroad', Wired.com. Online. Available at: www.wired.com/news/politics/0,1283,55530,00.html

Walker, L. (21 March 2003) 'Web use spikes on news of war', *Washington Post*. Online. Available at: www.bizreport.com/news/4248/

Walker, L. (26 March 2003) 'Casting a wider net for world news', *Washington Post*. Online. Available at: www.washingtonpost.com/ac2/wp-dyn/technology/columns/dotcom/archive?start=80&per=20

Walker, L. (9 May 2004) 'Iraq prison scandal at its most graphic', *Washington Post*. Online. Available at: http://electroniciraq.net/news/1799.shtml

Walls, A.T. (2002) 'The image of unanimity: The utility of the promotion and disparagement of cultural and social unanimity as a form of context manipulation in Information Warfare in the aftermath of the attacks of September 11, 2001', *Journal of Information Warfare*, 2(2): 119–127.

Wang, F. (undated) 'Subscribing to democracy through the internet', University of Pennsylvania. Online. Available at: http://mcel.pacificu.edu/jahc/jahcII3/ARTICLESII3/wang/wang.html

Waterman, P. (2001) *Globalization, Social Movements and the New Internationalisms*, London and New York: Continuum.

Weaver, N. (15 August 2001) 'Warhol worms: The potential for very fast Internet plagues', University of California, Berkeley. Online. Available at: www.cs.berkeley.edu/~nweaver/warhol.html

Webb, C. (17 January 2003) 'Overseas the Internet is rallying point for antiwar activists', *Washington Post*. Online. Available at: www.washingtonpost.com/wp-dyn/metro/specials/demonstrations/

Webb, C. (13 March 2003) 'Religious groups go online for peace', *Washington Post*. Online. Available at: www.findarticles.com/p/articles/mi_m0NTQ/is_2003_March_13/ai_98709639

Webb, C. (14 March 2003) 'Rallying around the flag online', *Washington Post*. Online. Available at: http://findarticles.com/p/articles/mi_m0NTQ/is_2003_March_14/ai_9877038

Webb, K. *et al.* (1983) 'Etiology and outcomes of protest: New European perspectives', *American Behavior Science*, 26: 311–331.

Webster, F. (1995) Theories of the Information Society, London and New York: Routledge.

Weinstein, L. (16 December 2002) 'Filters, laws won't clean up net', Wired.com. Online. Available at: www.wired.com/news/politics/0,1283,56855,00.html

Weiss, J. (23 July 2003) 'Blogs shake the political discourse', *Globe News*. Online. Available at: www.eightlinks.com/archives/000673.html

Wellman, B. *et al.* (1996) 'Computer networks as social networks: Collaborative work, telework, and virtual community', *Annual Review of Sociology*, 22: 213–238.

Wenzel, E. (18 November 2003) 'More privacy protections needed', Associated Press. Online.

Whine, M. (2000) 'Far right extremists on the internet', in Thomas, D. and Loader, B. (eds), *Cybercrime: Law Enforcement and Surveillance in the Information Age*, London: Routledge.

White, C. (12 June 2003) 'Iraqi blogger revealed', dot journalism. Online. Available at: www.journalism.co.uk/news/story662.shtml

White, C. (27 June 2003) 'China is top of the gaggers', journalism.co.uk. Online. Available at: www.journalism.co.uk/news/story673.shtml

Whittier, N. (October 1997) 'Political generations, micro-cohorts and the transformations of social movements', *American Sociological Review*, 62(5): 760–778.

Wilhelm, A. (2000) *Democracy in the Digital Age*, New York and London: Routledge.

Wilkinson, D.H. (27 May 2002) 'The web and conflict theory', geocities.com. Online. Available at: www.geocities.com/ResearchTriangle/1214/soc1.html

Wired.com (26 June 2002) 'Vietnam to monitor "reactionary" web use'. Online. Available at: www.guardian.co.uk/freespeech/article/0,2763,861190,00.html

Wired.com (29 June 2002) 'Fighting the network war', Wired.com. Online. Available at: www.wired.com/wired/archive/9.12/netwar_pr.html

Wired.com (9 December 2003) 'Kofi Annan: Keep Media Free!' Online. Available at: www.wired.com/news/politics/0,1283,61530,00.html

Wired.com (24 February 2004) 'Net dissidents jailed in China'. Online. Available at: www.wired.com/news/politics/0,1283,62391,00.html

Wired News Report (17 August 2003) 'Microsoft braces for blaster', Wired.com. Online. Available at: www.wired.com/news/technology/infostructure/0,60060-0.html

Wired News Report (23 October 2002) 'DoS attack maims web servers', Wired.com. Online. Available at: www.wired.com/news/business/0,55960-0.html

Witt, L. (16 April 2004) 'Finding truth on the internet', Wired.com. Online. Available at: www.wired.com/news/politics/0,1283,64967,00.html

Witt, L. (2 July 2004) 'Kerry net strategy now on voters', Wired.com. Online. Available at: www.wired.com/news/politics/0,1283,64066,00.html

Wo-Lamp Lam, W. (9 November 2003) 'China pressured over net dissident', CNN. Online. Available at: http://edition.cnn.com/2003/WORLD/asiapcf/east/11/09/china.net/

Wolfsfeld, G. (1997) *Media and Political Conflict*, Cambridge: Cambridge University Press.

Worth, R. (13 March 2005) 'Jihadists take stand on the web, and some say its defensive', *New York Times*. Online. Available at: http://siteinstitute.org/bin/articles.cgi?ID=inthe-news6605&Category=inthenews&Subcategory=0

Wright, S. (2004) 'Informing, communicating and ICTs in contemporary anti-capitalist movements' in Van de Donk, W., Loader, B., Nixon, P. and Rucht, D. (eds), *Cyber-protest: New Media, Citizens and Social Movements*, London and New York: Routledge.

Wright, S. (July 2001) 'Pondering information and communication in contemporary anti-capitalist movements', *The Commoner*. Online. Available at: http://libcom.org/library/information-communication-steve-wright

Yahoo (8 September 2003) 'MoveOn.org launches "recall no, democracy yes" campaign in California', yahoo.com. Online. Available at: www.findarticles.com/cf_dls/m4PRN/2003_Sept_3/107220680/p1/article.jhtml

YahooNews (29 October 2003) 'US cyber-diplomacy now up and running'. Online. Available at: www.infowarmonitor.net/modules.php?op=modload&name=News&file=article&sid=720

Young, D. (21 January 2003) 'Chinese chatting up with numbers in cyberspace', reuters.com. Online. Available at: www.clta-gny.org/numbers.htm

Zald, M. (1996) 'Culture, ideology and strategic framing' in McAdam, D., McCarthy, J. and Zald, M. (eds), *Comparative Perspectives on Social Movements: Political Opportunities, Mobilizing Structures, and Structural Framings*, Cambridge: Cambridge University Press.

Zerbisias, A. (1 March 2005) 'Why are we dumbing down the news?' commondreams.org. Online. Available at: www.commondreams.org/views05/0301-29.htm

Zetter, K. (5 September 2003) 'Security holes vex web host firm', Wired.com. Online. Available at: www.wired.com/news/business/0,1367,60303,00.html

Zetter, K. (11 September 2003) 'Just say no to viruses and worms', Wired.com. Online. Available at: www.wired.com/news/infostructure/0,1377,60391,00.html

Zetter, K. (5 May 2004) 'Blogs counter political plottings', Wired.com. Online. Available at: www.wired.com/news/politics/0,1283,63334,00.html

Zetter, K. (26 July 2004) 'MoveOn moves up in the world', Wired.com. Online. Available at: www.wired.com/news/politics/0,1283,64340,00.html

Zittrain, J. and Edelman, B. (14 January 2003) 'Documentation of internet filtering worldwide'. Online. Available at: http://cyber.law.harvard.edu/filtering

Zittrain, J. and Edelman, B. (accessed 20 March 2003) 'Empirical analysis of internet filtering in China', Berkman Center for Internet and Society, Harvard Law School.

Zucchino, D. (4 September 2004) 'Internet helps corral protesters', *Chicago Tribune*. Online. Available at: www.bushwatch.net/archive2004/e-mail-20040905.htm

Index